Bitter Wages

This book is printed on 100 percent recycled paper

Also Available:

Ralph Nader's Study Group Report on Disease and Injury on the Job

Grossman Publishers   New York   1973

Bitter Wages

by Joseph A. Page
and Mary-Win O'Brien

First published in 1973 by Grossman Publishers
625 Madison Avenue, New York, N.Y. 10022
Published simultaneously in Canada by
Fitzhenry and Whiteside, Ltd.
SBN 670–17048–8
Library of Congress Catalogue Card Number: 72–112512
Printed in U.S.A.

All royalties from the sale of this book will be given to the Center for Study
of Responsive Law, the organization established by Ralph Nader to conduct
research into abuses of the public interest by business and governmental
groups. Contributions to further this work are tax deductible and may be
sent to the Center at P.O.B. 19367, Washington, D.C. 20036.

The Study Group

Joseph A. Page, Co-Author
   Associate Professor, Georgetown University Law Center
Mary-Win O'Brien, Co-Author
   Third-Year Student, Georgetown University Law Center
Gary B. Sellers, *Project Director (until December 31, 1971)*
Ralph Nader, *Special Consultant*

Lynn Eden
Katherine Stone

Larry Bedard
Michael Charney
Gary Horlick
Brian Hurevitz
James Phelan
David Ransohoff
Richard Scheck
Robert Sears
Isadora Wecksler
Ben Williams
Anthony Young

SPECIAL PROJECTS:
   Mary Ellen Brennan
   Wyanne Bunyan
   Randall C. Stephens
CONSULTANTS:
   Andrea Hricko
   Donald Whorton, M.D.
   Sidney Wolfe, M.D.
PRODUCTION DIRECTORS:
   Ruth C. Fort
   Connie Jo Smith

This book is dedicated to the memory of Alice Hamilton, M.D. (1869–1970), pioneer practitioner of occupational medicine, outspoken advocate of the elimination of poisons from the industrial environment, feminist, and public citizen.

Introduction by Ralph Nader

In an address before the National Safety Congress, sponsored by the National Safety Council in Chicago four years ago, I discussed six impediments to any significant progress toward reducing the epidemic of occupational trauma and disease that was sweeping through the nation, then as now, with calendar regularity. They were (1) the widespread and very substantial underreporting by industrial firms of actual injuries and diseases incurred on the job; (2) the indentured status of plant physicians and safety engineers, with their inhibitions against candid diagnosis and reporting in professional journals; (3) the porosity of state safety laws and regulations, funded with an average budget of forty cents per worker per year, with little enforcement for recognized violations and less recognition for many occupational diseases as being on-the-job hazards; (4) the consistent lack of interest directed by union leaders to rank-and-file job hazards and the low priority accorded this issue in collective bargaining and the allocation of union funds and energy; (5) the anemic and secretive posture of federal efforts in research and advocacy for occupational health and safety under then existing laws; and (6) the creakingly obsolete system of workmen's compensation—from coverage, to benefits, to deterrence.

Not all these obstacles have remained unbudged in the past four years. Black lung disease has been officially recognized and is now compensable—more than thirty years after Great Britain acted. As a precursor of rank-and-file turbulence over long dormant but presently revealed occupational diseases, the black lung movement* is a signal to labor leaders that workers are increasingly capable of thrusting their own issues, independent of their unresponsive chiefs, into the public arena. This capability can be enhanced by the use of the federal Occupational Safety and Health Act of 1970. The first such comprehensive federal legislation, this law provides some tools that workers can employ to generate pressure at the job site for enforcement of safety and health standards and for the faster development of adequate enforcement levels.

Whether the decisive combination of factors can be brought together to make full use of the promise of the new law is the burden of this second historic stage of action on occupational casualties. The first stage emerged early in the century after the passage of state safety and workmen's compensation acts. The customary corporate cries of intolerable cost rang customarily phony as productivity increased and the deterrent impact of internalizing some of these job injury costs made good social, as well as economic, sense. But technology, the evolution of business organizations, and the atrophy of enforcement energy changed much faster than the occupational safety laws. The silent violence of occupational diseases expanded rapidly, while the thrust of the law was mostly limited to traumatic injury on the job. With budgets, data collection, inspections, and enforcement wallowing in an administrative miasma, the initiating role of government shifted by default to private constituencies that were barely interested. Both labor unions and insurance companies— the ideal countervailing powers to industry negligence—failed to make any significant headway on this major problem. They relegated it to a low priority. Voiceless victims individually afflicted do not usually produce the bread-and-butter issues on which labor leaders negotiate. Moreover, questions of occupational health and safety were never propelled into the political, economic, scientific, and communications arenas, where they could not have been so comprehensively ignored. They were al-

* Because other books and materials have covered the story of coal mine casualties and diseases, this book does not cover these problems in any detail.

ways a shadow over the land, a source of squibs for newspaper fillers, a crustacean attached to the hull of a national "that's-the-way-it-goes" accident syndrome.

The history of job safety policy and practice has many lessons to teach any future effort. It merits a brief summary. Before any legal or technical resurgence develops, there must be a broad sensitivity among the citizenry to this massive, continuing destruction of workers' bodies. As a form of violence, job casualties are statistically at least three times more serious than street crime, and with each new discovery or documentation of a hitherto neglected exposure to gases, chemicals, particulates, radiation, or noise, the epidemic looms larger and more pervasive. Most societies react to anthropomorphic, direct, and willful forms of violence far more seriously than to violent environments that are institutionally bred, indifferently allocated among the population, and associated with beneficial activities such as manufacturing or mining. The great psychological leap that has to be made to provide the public metabolism for enduring action is to realize that violence comes in many styles and does not depend on its motivation for its impact. Once this is realized, the problem can be properly grasped as a major challenge to our political-economic system's distribution of power and resources.

In some ways, the corporate attitude toward worker safety is similar to that toward pollution. As long as it costs less to permit casualties than to prevent them, the motivational direction of the company is not to invest in new equipment and procedures. The public interest, then, should call for the shifting of such costs onto the company, which is in the best position to detect and forestall job hazards at their inception and subject them to a relatively efficient degree of administrative control. That the burden of prevention should fall on industry is all the more important in the area of occupational disease, whose costs are now so deferred that their emergence in the form of emphysema, cancer, etc., is often at the moment of the worker's retirement or declining utility to the firm. This means that an adequate public policy must focus on risks as well as costs and, indeed, should treat risks as if they were present costs.

There is a more graphic similarity between worker diseases and pollution—they often are different manifestations of the same industrial process. The contamination of the air, land, and water outside, for example, is often worse inside the plant or mine. On-the-job exposure to pollution—such as coal dust, cotton dust, lead, carbon monoxide, mercury, beryllium, asbestos, dyes, and pesticides—is very often many times more dense or intense than in the general environment within the particular pollution zone. It is time to diagnose medically that ultimate depository of environmental contaminants—the human being—and take the cues for action from that basic touchstone of our proper values. For we are dealing with an affliction whose seriousness the injured often do not feel or understand until it reaches irreversibility. Men and women absorbing industrial contaminants into their bloodstream or lungs do not know the consequences. They know little of how preventable these pathologies are—through available or knowable technology and work processes. They don't realize that jobs and hazards don't have to come hand in hand.

The complete line of defense or offense should come from the unions—both on behalf of their members and their unorganized brethren. Unfortunately, it has not. Union leaders, with their swollen treasuries and shrunken imaginations, have almost uniformly failed to equip their staffs with the skills to locate and detect the full range of job hazards and to develop strategies for prevention. Most of the large unions have few or no safety engineers, industrial hygienists, physicians, or lawyers specializing in advancing the workers' right to a safe working environment. The United Mine Workers leadership disgraced itself on the issue of black lung disease that was eroding the health of half of its members. The International Brotherhood of Teamsters has nobody in Washington working full-time to achieve safer working conditions and safety equipment for truck drivers. The United Auto Workers has only four full-time health and safety staff members at its national headquarters (mostly working on compensation problems), despite highly harmful job hazards to thousands of UAW workers in auto plants and foundries. Despite the desperate need for the Departments of Labor and of Health, Education and Welfare to monitor the administration of the new federal law, the main labor unions view a skilled, full-time staff working to pressure the Departments toward this objective with a shocking combination of myopia and intransigence. Job hazards are only rarely a major issue in labor-management negotiations.

Much of the secrecy, nonenforcement, and industry-indentured practices of state and federal regulatory agencies described in this book could never have prevailed if union leadership had been alert and actuated by the needs of its members. It is of course true that other private constituencies could also be doing much more. Safety engineering societies, medical societies, workmen's compensation lawyers, and, especially, the powerful insurance companies and their trade associations come to mind. But all had little to gain by breaking up the comfortable occupational safety establishment, which served to camouflage the shame of the industrial and commercial moguls with false standards, prizes, medals, self-censorship, and noncontroversial symposia.

The willingness and power of this establishment to corrupt or restrain is nowhere more dismaying than in the case of the plant physician. Dr. Lorin E. Kerr, the United Mine Workers' only full-time physician on miners' health, had this to say a few years ago about what he described as "the attitude and influence of employer-oriented physicians who avoid facing known facts about the ravages of coal dust in human lungs."

> The impact of physician knowledge and attitudes is apparent in inaccurate or incomplete death certificates. These documents are essential for determining the existence of a dangerous dust requiring governmental action for control or eradication. Too often no records mean no action. Physician denial or avoidance of known facts can also determine the publication of information. In 1956 the concern of some physicians about management attitudes effectively blocked the publication of a paper of mine on coal workers' pneumoconiosis in the most widely circulated medical journal in the United States. One can only wonder how many times other medical information vital for the protection of the health of workers has not been published for the same reason. Physician attitudes also influence other groups such as company lawyers, state legislators, and officials who, all told, are remarkably resistant to the elimination or control of coal dust because it would cost money.

Since publication of clinical information about job-concerned diseases in the medical journals is often the condition precedent for follow-up action, this pattern of censorship has nipped many changes in the bud. A comparison of medical literature in Britain and the United States over the years on occupational diseases shows much earlier publication and much greater professional independence in the British journals.

Everywhere one turns, the wreckage of the hopes of the early twentieth century advocates of job safety is evident. Companies routinely violate with impunity state and federal safety laws and avoid even the obligation of reporting their workers' injuries. West Virginia officials, for example, admit this widespread reporting violation but have done nothing about it. Texas, on the other hand, didn't bother to pass an occupational health and safety law until 1967 for its 3.5 million workers. Companies routinely play one state against another to the lowest common denominator of safety legislation and enforcement by threatening to move to another state that is even more lenient. And the anarchy that prevails over the shocking conditions for migrant and other agricultural workers throughout the country is almost beyond description.

It is quite likely that the rising dissatisfaction of workers over the tedium, repetitiveness, speed-up, and mind-numbing nature of industrial job organization—embodied by the 1972 strike at GM's Vega plant at Lordstown, Ohio—will compel more attention by the nation over just what labor conditions are like on the job. This greater focus, first involving union locals, should launch a comprehensive awareness of the inseparability of occupational stress from mental health and the status of being human in a society whose wealth and technology could permit so much more self-development and fulfillment, as well as safety.

The American attitude toward work has consistently been that it is something to do so that the rest of life can be more bearable or comfortable. Work has had little meaning in and of itself except in the professional or craft areas. With assembly lines and automated processes, it was easier for the worker just to go through the motions without feeling a part of anything that counted or that was not expendable. Finding meaning in work is more than a quest for enjoying that part of the day on the job; it is also a way to develop a sense of normative participation in what the work is all about within a broader scale of human values—starting with company decisions relating to matters such as pollution, job hazards, consumer abuse, and compliance with the law. In this wider context, the contemporary, second-stage drive for job health and safety is part of a larger struggle for workers to share in the design and operation of the work space. After all, management mistakes or callousness are visited on the workers in the most intimate invasion of their physical and psychological integrity. Merely compensating for such injuries or diseases is the sign of a catch-up, not a humane, society.

This book, with its disclosures, analysis, and recommendations, may help accelerate the support given those few stalwart persons in the labor movement and in the professions who have been striving to move their sluggish organizations toward response. As much as anything else, the determined striving of those individuals heartened the medical, law, and engineering students, and young lawyers who worked to prepare this study.

Ralph Nader
April, 1972
Washington, D.C.

Contents

Bitter Wages

1

Voice from the Scrap Heap

Allen Cooper of Detroit, Michigan, is an articulate, pleasant, and soft-spoken man who, at the age of forty-nine, is blind and has trouble breathing.

Cooper grew up on a farm in Red Spring, North Carolina, and went as far as the eighth grade when he quit school in order to work. He moved to Washington, D. C., in 1945 and stayed there until 1949, when he journeyed to Detroit to take a job with Midland Steel, a corporation that manufactured parts for the automobile industry. He was twenty-seven years old at the time, and in good health. Cooper told one Study Group member about his experiences at Midland Steel as a bar puller, welder, painter, steel grinder, and general repairman.

"I was a bar puller and that was near a lot of paint fumes and steam, lots of steam mixed up with paint fumes. The bar we pulled was the bar that hooked on the front end of the frame [of the automobile]."

On this job Cooper lifted bars weighing seventy-five to ninety pounds, and the bars were hot. "You're taking bars off the frame as it goes through the washer, dryer, then spray paint, then heat treatment, which is 150, 160 degrees, or maybe it's 200. Then when they come out to you pretty hot they come on a roller table and you had to pull it, take the hooks out on the side.

"You had two pairs of gloves. They were heavy leather with rivets in them and a cloth pair on the inside because those rivets got hot. But your hands got hard as nails. It was very hot. To tell you the truth, you got so that when you weren't working, you were still working. Really, it was a man-killing job, pulling bars. I pulled them for about two, three years.

"When I quit that, I started welding. And that wasn't so efficient, you know. Lots of times you'd have to step out of the booth to try to get some fresh air for a couple minutes. I welded steady for about two years. I was welding frames, auto frames, putting the last hanger on the rail.

"It was arc welding, electric arc welding. I did a little gas welding; I think the electric was worse. The gas welding was out in the open, we were repairing frames. The electric was worse. It was mostly on the line and there were so many [men] welding. There was a guy right next, and right across from you. And if you didn't do your job, the supervisor was on you. It kept you on your toes.

"I worked all over the place. I was a spray painter and worked in the dip tank." In the dip tank department Cooper lifted rails and truck frames out of a tank of paint and solvent. He had paint and solvent splashed into his eyes a number of times, requiring first-aid treatment. "You had a little hoist to let it [a rail or truck frame] down. First it was washed in detergent." These detergents are caustic and coagulate protein—whether in the eye, skin, or respiratory mucosal lining. "It was hot enough to burn you. You had to hook it and sometimes it was not six inches from your eyes. You couldn't afford to wear goggles because you could hardly see anyway. The fumes were the worst here, because of all the steam. It was worse than bar pulling. There were fumes there, but no steam. The dip tank department was just full of steam. There were doors to let it out, but they didn't open them because then the rest of the factory would be steamy. The steam was just in one hundred-foot-wide room, a straight long room. There was no way to take the fumes out. You could taste them in your throat all the time." At that time, when Cooper coughed, his sputum was black.

"Did the union do anything about health? No indeed. Somehow I think the union went along with the company. Did they do anything about safety? No, no they didn't. As I say, at that time Midland Steel paid a little more than anyone else around and one man complaining didn't do any good."

Cooper then requested a job change from the dip department and was assigned to spray painting. The parts he painted were moved by gasoline-powered tractors that also gave off strong fumes and exhaust. Cooper did not wear goggles or a mask, although the area was filled with fumes from the paint and solvents. He worked this job over half a year.

Allen Cooper also hooked steel in the pickling room (one step in processing steel). The pickling tubs contained acid and emitted fumes that were irritating to his eyes, nose, throat, and chest. The fumes cut his breath, he said, and he had to gulp fresh air for relief. While working in the pickling department, he had severe coughing and shortness of breath and coughed up blood many times.

Cooper ended up welding again. "Was there noise? Oh yes. We had some big presses in there, twelve, fourteen, sixteen feet long. And when you cut that steel the sound was like thunder. I know quite a few people that have taken sick. If it wasn't the eyes, it was the mouth, their teeth rotted out. And then there's quite a few heart conditions.

"Ever since I started working there I had trouble with my eyes. Well, not right away, but in two, three years. My eyes began to run water and they kept getting worse and worse, you know. Then when I left there it was the same thing. I didn't have no dreams of compensation when I left."

Cooper became aware of decreasing vision in about 1955, when he worked as a welder and found himself suffering from welding flashes. He had to return to the plant first-aid department from his home at night because of pain in his eyes. He was given drops to put in his eyes to ease the pain and was often sent back to work with his eyes still moist from the drops. These eye drops are especially dangerous to workers because they usually contain local anesthetics; hence, the worker is unable to tell if new particles or solvents have entered his eyes, and he is made especially vulnerable to new injury. Cooper's work record from late 1956 to the middle of 1957 shows he suffered injury to the coating of his eyes four times because of welding flashes.

Because business was bad, Cooper was laid off in 1958. He was thirty-six years old at the time and had been on the job for only nine years. "I didn't get a bonus even. They wouldn't sign the last month lay-off slip. I went out there and they told me, 'That's okay, we know you've been here, we don't have to sign it.' But then they wouldn't give me that money.

"After I was laid off I worked in the Mayflower Donut Shop in Washington, D.C. I was head porter there and was sweeping and cleaning the floor with a detergent that was not strong, but I just started getting weak. They let me get off at Christmas and I came home and thought I was feeling pretty good."

Then Cooper said, "I woke up one morning and couldn't see. When I went to the hospital they thought that I had sarcoid. Sometimes there were conferences of fifteen doctors." Cooper also had welts across his back: "I'd break out in whips across my chest, my back. It was like ropes. Leaves your skin raw, scaly, and rough. I was in the hospital in '60. I was there straight two years.

"But then I couldn't work again. I applied for compensation and they give me fifteen hundred dollars. 'That's all I can get,' they said. 'I'm not able to hold another job,' I said. 'Yeah,' they said, 'that's all you can get. You have fifteen days to decide.' And before the fifteen days were up they had sent the check to me. I said they seemed really anxious to send me money. Well, my wife said, 'Why not take it, you might not get anything otherwise.' And well, I had the check, and why not? So I kept it. Lately I was talking to my minister and he said maybe it could get re-opened, the case."

The Study Group member asked Mr. Cooper how much he can see.

"Very, very little," he said. "There's a little peephole in my right eye. My left eye is gone."

Can he see color?

"If it isn't too much pressure I can. Otherwise I can't; I just see shadow. As long as I take my pressure pills and don't try to do too much I'm okay. That pressure is terrible. The channels are clouded up. I take medication such as Diamox, eyedrops, special drops for my right eye, and special drops for my left eye.

I'm going in next week for a checkup. I was getting social security, but social security is so slender you can't do very much with it."

How does he feel?

"I don't feel so good. Sometimes I feel all right. Sometimes I feel weak. And when I feel weak the pressure rises. Then I just have to lay out and keep low. Sometimes I fall asleep and wake up and feel a little better. The pressure is from my ears all the way around my head. In my right ear, they found out I wasn't hearing anything from it."

Cooper's doctor, Dr. Janette Sherman, confirmed Cooper's analysis of the connection between his work and his bad health. She found that he has marked decreased pulmonary capacity and that his cough now produces thick white sputum, which is occasionally blood-stained. (As for the diagnosis of sarcoidosis, Cooper observed that the doctors who treated him years ago told him sarcoidosis "came from pine trees. But I never did mess around no pine trees much, but I messed around thinner.") While one theory of the origin of sarcoidosis attributes its specific inciting agent to pine pollen, no definite relationship has been established, and it is most accurate to say that the cause of sarcoidosis is unknown. Pulmonary involvement is the most common and perhaps the most serious manifestation of sarcoidosis, which also may produce visual impairment and blindness.

Dr. Sherman did not know whether the precise diagnosis of sarcoidosis some years before was correct or not. She said that even if Cooper did have sarcoidosis, his job could easily have adversely affected that condition. She also said that a diagnosis of sarcoidosis in no way means that his condition was *not* caused or aggravated by his work. Dr. Sherman said that "sarcoid" is merely a name given to a constellation of symptoms that may include lung and/or eye lesions. She also pointed out that the lesions of sarcoid look exactly like those of beryllium, which is used in making steel alloys. While Dr. Sherman was not saying that Cooper had berylliosis, she thought that there was some room to doubt the validity of the original diagnosis since beryllium is used in making steel and could have been vaporized by the welding operation and been rendered as particulate matter by the other milling processes. She felt that from Cooper's work record, no one could fairly conclude that his condition was unrelated to his work.

The Study Group member asked Cooper whether he felt any anger about his experience.

"Yes I do, but I haven't found anyone who could do anything about it. I told my lawyer he could have done more than he did. He told me it's been so long—but they have it in black and white. My lawyer said sarcoid could come from so many different things. I said I know myself and I know where it began. He said that's the best he could do for me. He got me fifteen hundred dollars and took five hundred of that. I couldn't understand it. I feel it's a made-up thing between the referee and the lawyer.

"I know what I've been through. They want to get every drop of blood out of you. Let me ask you a question now. Do you think I'm right? Can the case be reopened?"

Allen Cooper is but one tiny speck on a human scrap heap created by the American industrial juggernaut. With him on the pile are the corpses of men and women killed on the job, more than fourteen thousand of them each year; the millions of workers disabled annually from work-related accidents; and those numberless individuals who, like Allen Cooper, have fallen prey to the ravages of job stresses and exposures.

Industrial mishaps have produced annual fatalities that have exceeded war deaths in Vietnam[1] and Korea[2] during comparable periods. As Ralph Nader remarked to a Senate subcommittee in 1970, "In the last three years the fatality toll of riots in our cities has been 260 to 270 dead. Just in total fatalities . . . that amounts to five days' toll in the occupational safety area."[3]

Industrial diseases take a horrible toll of the work force. Accurate statistics reflecting the incidence of recognized job illnesses simply do not exist. In 1968 the U. S. Public Health Service cited the figure of 336,000 cases annually, but this was nothing more than a national projection of the 27,000 cases reported in 1965 in California, the only state in the union that has a comprehensive system of recording occupational diseases.[4] Thus, the statistic did not even begin to encompass diseases related to types of work not found in California, such as coal miners' black lung, which afflicts one in ten active miners and one in five former miners,[5] and kills 1,000 miners each year in Pennsylvania alone.[6]

The National Institute for Occupational Health and Safety came up with some new estimates in a 1972 report: 390,000 cases of job-related illnesses, and 100,000 deaths from industrial diseases.[7]

In 1968 a team of U. S. Public Health Service industrial hygienists conducted a walk-through survey of the work environment in over eight hundred randomly selected plants in and around Chicago. They found that 70.7 percent of all benzene exposures were poorly controlled, as were 56 percent of exposures to carbon monoxide, 81.8 percent of exposures to carbon tetrachloride, 67.9 percent of exposures to lead, and 57.9 percent of exposures to free silica.[8] An analysis of the study concluded that a "projection of the 1968 survey findings suggest [sic] that more than one-third of the 1,048,851 in-plant workers [in the Chicago area] are exposed to potentially hazardous working conditions and that approximately 63 percent of the exposures are poorly controlled."[9]

Dr. William H. Stewart, the former U. S. Surgeon General, stated in 1968 that:

> In 1966–67, we studied six metropolitan areas, examined 1,700 industrial plants which employed 142,000 workers. Because of the precise nature of the analyses, it can be statistically related to 30,000 plants, covering 650,000 workers. The study found that 65 percent of the people were potentially exposed to toxic materials or harmful physical agents, such as severe noise or vibration.
>
> Our investigators examined controls that were in effect to protect these workers from toxic agents and found that only 25 percent of the workers were protected adequately. The remaining workers were plainly unprotected or working in conditions which needed immediate attention.
>
> Recently, we studied foundries in one state. We found that 7.3 percent of the 3,200 workers—or one in fifteen— were exposed to environmental conditions which were capable of producing disabling and fatal diseases.[10]

In addition to the scant available data on industrial diseases, medical and scientific studies increasingly fortify the supposition that many workers disabled by diseases not thought to be occupational in origin are in fact victims of job-related illnesses, and that many workers seemingly sound in mind and body are undergoing constant, insidious exposures to industrial stresses and substances that are destroying their health.

Modern technology, moreover, has been spawning health hazards at a dizzy pace, and the nightmarish brood continues to multiply. New, potentially noxious chemicals find their way into industrial use at an estimated rate of one every twenty minutes.[11]

Despite these grim facts, a ray of hope flickers from the growing awareness on the part of workers, like Allen Cooper, that they are being sacrificed on the altar of the industrial environment. Appalachian coal miners, once they are made aware of the facts, have displayed a militant resistance to the corporate and governmental neglect that has consigned them to the agonies of black lung disease.[12] In a study sponsored by the Department of Labor and conducted by the University of Michigan's Survey Research Center, 71 percent of a national sampling of workers rated work injury or illness as a "very important problem," the highest percentage registered on a list of problem areas that included salary, fringe benefits, and unsteady employment.[13]

Recent statements by members of the Oil, Chemical and Atomic Workers International Union illustrate the dimensions of this awakening consciousness and eagerness to learn. Peter MacIntyre, president of the Chemical Workers local in Sayreville, New Jersey, asked his national union:

> What can be done, what can be told about gases when they're mixed together, such as chlorine and titanium tetrachloride? Now we have operators who have been working with these gases and fumes since 1961. We'd like to know what's happening to these people. Some of them have been taken out of the plant. Some of them, nothing has ever happened to them. We would like to know would wearing clothes that smell from chlorine be dangerous? We have people who continuously have their clothes saturated with fumes of chlorine.[14]

Another worker at a chemical plant in Linden, New Jersey, asked the national union about the effects of acrylamide:

> We've had six or seven people that have suffered strokes, paralysis. One of the men became blind about a year ago. Now this acrylamide is also used in all acrylic-based

paints. . . . But what I'm interested in now is finding out exactly what the crippling effects of this acrylamide are. Because everybody in this plant is exposed to this, due to the faulty equipment that management has installed there. They're only concerned with a production yield, not a safety standard.[15]

Perhaps the most touching testimony came from Harold Smith, who works in a plant where workers are exposed to mercury and chlorine:

And, well, my father worked in a chemical plant right next door to the one I work for; about twenty years. He's dead now. I had an uncle; he also worked in a chemical plant, the same plant right next door to me. He died of cancer, this cancer in the throat. He had a tube in his throat, and it was as a result of working in this chemical plant; he didn't have it before he went there. But a certain chemical that he inhaled got in his throat, and his throat was a mess and he died. I mean, I don't use the expression—he died like a dog. . . . We're a small bunch but we've got a problem. These chemicals are going to kill us all.[16]

In the final analysis, it is the worker himself who is most vitally touched by the conditions under which he labors. Since the dawn of the Industrial Revolution he has learned to live with the traumatic job accident—the slip and fall, the fire, the explosion, and the unhappy tendency of machines to mangle human limbs—and to accept these as an unavoidable part of his existence. Only recently has he begun to raise serious questions about this inevitability, and to grow aware of the extent to which occupational diseases are affecting him.

This book will seek to fortify this burgeoning consciousness. Succeeding chapters describe the principal known health hazards that have turned the workplace into an arena of invisible violence; the long history of governmental attempts to deal with job safety and health; the roles played by the states, federal agencies, and the labor unions; the passage of the Occupational Safety and Health Act of 1970; and its subsequent implementation. Our goal is to explain why past efforts have failed to provide the American worker with a reasonably and justly safe and healthful job environment, and to suggest what should be done to achieve this end.

The Silent Violence

In the Haskell division, where they work primarily with the metallic pigments and powders . . . there is a constant haze of dust. . . .

If you have seen the Wizard of Oz, there is a character called the Tin Man. Every man who comes out of there looks like the Tin Man. His eyes, nose, face, clothes. It is in his pores.

— *Testimony of Lou Laplaca, president, Local 1668, United Auto Workers, before the Senate Subcommittee on Labor, March 7, 1970*[1]

It may be that the dramatic interaction of man and industrial society is playing itself out most fully in the human lungs, the main gateway to the body tissues and a vital organ particularly vulnerable to the work environment. The lungs are one of the more intricate and beautiful structures of the universe, as befits their complex function of constantly providing the body tissues with oxygen, and removing the waste product, carbon dioxide.

The lungs are highly elastic, each enveloped in a double membrane called the pleura, which covers the outside surface of the lung and the inside surface of the chest wall. These two layers form a closed cavity and are separated only by a thin film of fluid, which enables them to glide with very little friction over each other as the lung expands and contracts in breathing.

Air enters through the trachea, or windpipe, which divides into two bronchial tubes. Each bronchial tube enters a lung and then further divides, and then subdivides again and again, distributing itself throughout the lung like the branches and twigs of an upside-down tree, until the finest tubes, the bronchioles, or capillary bronchi, open out into tiny air sacs called alveoli.

A complex of extremely delicate membranes, the alveoli collectively form a lacy, globular structure with a surface equivalent to the area of two tennis courts. They are surrounded by an extremely dense network of blood capillaries, which have likewise branched off from the pulmonary artery that comes from the heart. The air in the alveoli is separated from the bloodstream by two delicate membranes, the wall of the alveoli and the capillary wall, through which oxygen readily passes into the bloodstream and carbon dioxide is expelled.

The lungs contain a very efficient cleaning device. Some gases and most large particles are scrubbed out from the respiratory tract as far down as the terminal bronchioles by the action of tiny, hair-like cilia, which vibrate and thereby drive a thin layer of mucus, secreted by numerous glands of the bronchial tubes, up toward the pharynx, or throat, to be eliminated from the respiratory system.

Efficient though this system may be, it cannot expel all of the gases and fumes that may be inhaled in great concentrations, nor can it trap small dust particles before they reach the inner recesses of the lung. Dust particles between one fifty-thousandth of an inch and one five-thousandth of an inch in diameter commonly elude the cilia of the lung and penetrate to the terminal bronchi and alveoli, where they remain. Many of these dusts, particularly those with chemical properties, have a latency period that postpones the recognition of their serious effects on the lung or body until long after their original exposure.

Thus, the respiratory system presents a quick and direct avenue of entry for toxic materials into the body. The accumulation of these materials can lead to serious ailments such as pneumonia, tuberculosis, and bronchitis, which can be treated, and emphysema, a chronic, obstructive, and incurable lung disease.

In 1968 the U. S. Surgeon General reported that "the Social Security Administration is currently receiving about thirty-five thousand claims each year for disability from emphysema. About seven thousand claims—or one-fifth—are originating from occupations where the emphysema rate is excessive and the occupations are known to involve exposure to materials which are damaging to the respiratory tract."[2]

The lungs of an emphysema victim are overdistended with air that cannot be readily expelled. The walls of the air sacs in the lungs stretch and tear. As the process continues, the trapped air enlarges the lungs and causes a struggle for breath. A victim of emphysema cannot continue to work normally and often becomes a pulmonary cripple. His entire life centers around the task of breathing. Very often he cannot sleep unless he has several pillows or is sitting in a chair. Even walking half a block may be too much exertion. Emphysema causes strain on the heart and often leads to death from heart failure, as the heart eventually tires out from the markedly increased strain. Emphysema also weakens the lungs' defenses and makes them more vulnerable to infectious respiratory diseases as well as a number of other illnesses. There is no known cure for emphysema at this time and the damage it inflicts upon the body is irreversible.

No statistics are available to show how many American workers are affected by various dusts and particles. From 1959 to 1962, the only period for which the federal government collected complete records of recipients of social security disability awards, more than six thousand males (none of whom were miners, and all of whom were under sixty-five years old) were awarded payments for permanent *total* disability resulting from occupational dust diseases.[3]

In 1967, the three employment categories with the largest numbers of workers exposed to dusts contained over 3 million employees—1.3 million in industries producing primary metals, the same number in industries producing fabricated metals, and 629,000 workers producing stone, clay, and glass.[4] What follows is a description of some of the dusts these people inhale and some of the diseases they consequently endure.

Silicosis

> *It* [*the Congressional committee investigating the Gauley Bridge disaster*] *has heard a mother, whose eyes were those of a stricken animal, tell of the slow strangulation within the course of thirteen months of her three young and vigorous sons, and of the youngest who asked that he be "cut open after he was dead so the doctors could find out what had killed him."*
>
> —*Vito Marcantonio, "Dusty Death,"*
> The New Republic, *March 5, 1936*

The most prevalent of the dust diseases (which are collectively known as pneumoconioses) is silicosis. Silicosis is caused by the inhalation of particulate matter containing free silica, or silicon dioxide. It probably existed as far back as the Paleolithic period of man, when flint implements were first used. Hippocrates and Pliny recognized its symptoms. The first account of the pathology of silicosis was written in 1672 by a Dutchman, Isbrand van Diemerbroeck, professor of medicine in Utrecht, who described the lungs of several stonecutters who had died of "asthma." When performing autopsies, he found that cutting their lungs was like cutting a mass of sand.[5]

Silicosis has continued to take the lives of workers. Since the beginning of the twentieth century, it has accelerated in severity as a result of the mechanization of industry and the increase in the amounts of dust generated by airhammers, spray guns, and other automatic tools.

Free silica can be found in quartz and other rock. It endangers sandstone workers, sand blasters, metal grinders, and tin and coal miners, as well as workers in foundries and pottery factories.* Silicosis is a worldwide problem and, of all the pneumoconioses, it claims the largest number of victims, either by itself or with tuberculosis, an often-recorded combination. Like

---

* Silica in the nonfree or combined state, called a silicate, refers to silica in chemical combination and includes the feldspars, asbestos, kaolin, mica, shale, slate, talc, and turpentine. Pneumoconiosis associated with the inhalation of silicate dust is called silicatosis.

all the pneumoconioses, silicosis is marked by shortness of breath, coughing (with or without blood), difficulty in sleeping, chest pounding, chest pain, and weakened resistance to pneumonia, bronchitis, tuberculosis, and other lung diseases. All this leads to serious physical incapacity—such as the inability to climb stairs and to continue with usual activities—and to death.

Precisely how silicosis affects the lungs is unknown, but present theory is that the biological action of the silica crystal, when inhaled into the alveoli, sets in motion a tissue reaction causing numerous round scar nodules to form throughout the lung.

Black Lung

> *Coal dust was everywhere. Coal dust to Melonsville, Pennsylvania, is what sand is to the frontier part of Arizona. It lies in drifts against the foundations of the house, in ripples in the roadway. Every step sends it up to the eyes and nostrils in choking clouds. Every wind threshes it across the prospect. But in wet weather it is, if anything, even more disagreeable. Under the pelting of the rain at the end of that bleak day on that bleak hilltop it had resolved itself to mere sodden mud—mud that clung, and clung and stuck with the tenacity of oil, and streaked and stained everything it touched.*
>
> —*F. Norris, "Life in the Mining Region,"* Everybody's, *September, 1902*[6]

When the lungs of city dwellers are cut open at death, they are fairly black. When the lungs of miners suffering from coal miners' pneumoconiosis are cut open at death, they are extraordinarily black. Hence, the name "black lung." Like other debilitating lung diseases, black lung leads to shortness of breath, incapacity, and death.

It differs from silicosis in that the deposition of coal dust, rather than free silica, appears to be the responsible agent. Coal dust accumulates near the terminal bronchiole and forms small nodules called coal macules. The most frequent effect on the lung is focal emphysema, which happens when the alveoli behind the coal macules become disrupted. Focal emphysema is not necessarily as widespread throughout the lung as emphysema. Although focal emphysema causes marked functional disability, the chest X ray and simple breathing tests (spirogram) may show no abnormalities. Nevertheless the miners show marked shortness of breath from minimal exertion and are unable to continue working.

Kermit Clark, of Weeksberry, Kentucky, described to a Congressional committee a victim's-eye view of the symptoms of black lung:

> I was a coal miner for thirty-nine years. I went to work in the coal mine when I was thirteen years old. Out of the thirty-nine years, I served five years in the armed forces in World War II. I had to quit work on account of pneumoconiosis, or black lung, whichever you want to call it, and I was advised by ten different doctors to quit work, that I am not supposed even to drive my car. I am not supposed to go fishing. I am not supposed to do anything according to the depositions from the doctors I have, and I am supposed to fall dead any time. I never sleep at night. I stay awake and sleep on the average about two and one-half hours a day, and most of that is sitting in a chair. I never had any pleasure out of life at all since I had to quit the mine. On eight different occasions I was hauled out of the mine because I was passing out because I could not get enough air to breathe.[7]

Many miners die because of heart failure from the increased strain placed upon the heart by focal emphysema. A small percentage of miners (2 percent) develop massive fibrosis (or scarring), which destroys one or more major areas of the lungs and also causes a marked stress on the heart.

This century has seen a dramatic increase in the number of black lung victims as a result of the mechanization of the coal mines and the generation of more dust from fast-operating machines. Roughly 10 percent of all active coal miners have black lung.[8] In 1969 the U. S. Surgeon General estimated that close to one hundred thousand active and retired coal miners suffer from the disease.[9] Among coal miners the death rate from respiratory diseases is about five times that of the general working population.

Byssinosis

I have worked all of my life in the cotton mill . . . for forty-four years. I started to work when I was real young. Before I even finished grammar school, circumstances forced me to go to work.

When I went to work, I was in good health, I thought, as a boy of that age. I could get out and run and play and wasn't bothered any. So I went to work in the carding department. . . .

The mill then was much more open than today. But I began to notice when I would play, I couldn't breathe as good as I could before I went to work. Being young and not thinking much about it, I kept on going at it, and finally I married and then I was hooked to keep on working.

So I kept on working and my breathing kept on getting worse. I went along, and I got to the point where I would just cough and sometimes hang over a can and cough and become nauseated in my stomach, and I began to be bothered about my condition.

I decided that maybe this dust was doing that. I started taking a little bit closer watch on that and taking a little closer check on it. I have seen three-hundred-watt electric bulbs in the plant that I worked in where the ceiling was high, and they looked like they were red and not much more than a twenty-five-watt bulb. When the mill would run two or three days, you could look up and see the light bulbs and they looked like they were red. You could get over by the window where the sun was shining through, and the dust particles were so thick when you looked into that sunshine that it looked like you could just reach out and grab a handful of it. . . .

Apparently, people would look at me and say that I am in very good physical shape. All of my life I have been very active, but now I am short of breath, and I can't do

*anything. I get so short of breath and so weak, I can hardly
go.*

*As far as I see things, and as far as the doctor says, I
will be like this as long as I live.*

—*Testimony of Lacy Wright,
before the Senate Subcommittee
on Labor, April 28, 1970*[10]

Over seventeen thousand active cotton, flax, and hemp
workers suffer from byssinosis, or brown lung.[11] Among those
groups specifically exposed—those engaged in the initial processing of cotton yarn and in the initial stages of processing flax and
hemp fibers—more than 90 percent of the workers may have
byssinosis.[12] Although there was evidence of respiratory symptoms similar to byssinosis in the hand-picked cotton era, an increase
in the incidence of the disease has followed the mechanization of
the cotton industry in the 1950's.[13]

The initial symptoms of byssinosis are feelings of tightness
in the chest on Monday mornings after returning from a weekend off work. A leaflet put out by the Department of Health,
Education and Welfare (HEW) says:

> Breathing tests show a reduction in breathing capacity on
> Mondays, even among some workers who don't feel the
> chest tightness. In some mill workers—no one knows how
> many—this condition becomes steadily worse over the years.
> The "chest tightness" extends to Tuesdays, Wednesdays, and
> eventually throughout the week. In some workers the breath-
> ing capacity continues to drop until shortness of breath be-
> comes severely handicapping.[14]

Byssinosis, unlike other dust diseases, does not show up on
chest X rays, nor have pathological changes specific to this disease been identified in the lungs of workers suffering from byssinosis. Any chest X ray changes or lung pathology incurred by
these workers resemble those also found in chronic bronchitis or
emphysema from nonoccupational causes. The lack of this kind
of objective evidence has for a long time fortified the efforts of
industry groups to escape responsibility for the disease. How
strong these efforts have been can be seen from this excerpt
from *America's Textile Reporter*, a trade publication:

> We are particularly intrigued by the term "byssinosis" a
> thing thought up by venal doctors who attended last year's
> ILO [International Labor Organization] meetings in Africa,
> where inferior races are bound to be afflicted by new diseases
> more superior people defeated years ago.
>
> As a matter of fact, we referred to the "cotton fever"
> earlier, when we pointed out that a good chaw of B.L.
> dark would take care of it, or some snuff. . . .[15]

America's Textile Reporter to the contrary, the prevalence
of byssinosis has been established, in many countries, by epidemiological surveys and pulmonary function tests. The disease was
described in medical literature published in Great Britain in the
1930's. Basically, a certain part of the cotton boll, the bracts,
which are more widely encountered in the earlier parts of textile processing, contain a pharmacological agent that induces
broncho-constriction. This agent causes the lungs to release
histamine, causing constriction. The reason the symptoms are
most often felt on Monday is because the first exposure releases
the available histamine in lung tissue, leaving insufficient releasable histamine to cause symptoms on subsequent working days.
After a two-day rest on the weekend, the lungs have stored
sufficient releasable histamine to produce constriction again the
following Monday.

What is not understood is the cause of byssinosis—why the
presence of a histamine-releasing, broncho-constrictor agent in
cotton dust causes chronic, irreversible lung-function loss in textile workers. There can no longer be any doubt, however, that
this is just what it does.

The callous attitude of *America's Textile Reporter* com-
pares poorly with the sensible and humane approach followed in
Australia. According to Dr. Bryan Gandevia:

> We are a pretty simpleminded and unsophisticated mob down
> under. I do think that we don't seem to have quite the same
> difficulties to which your attention has been drawn during
> this conference. In Australia byssinosis has been a compen-
> sable disease by regulation for some years. . . . A disease does
> not exist in an industry. It exists in a society. It exists in an
> entire sociological framework. In looking at the disease we
> must look at it from a pretty broad point of view. In
> Australia it is accepted as an undesirable disease and this,

I think, helps to investigate it, to understand it, at least at a fairly empirical level. We don't as a rule have tremendous difficulty in recognizing it, whether by survey methods or in the individual case. We don't on the whole have tremendous difficulty in making practical recommendations regarding the continuing supervision of card-room and other employees in the cotton industry. We don't really have a tremendous amount of difficulty in making practical and reasonable suggestions as to the control and prevention of the disorder. One of the difficulties, I think, which one may detect among leaders of industry, is that you find certain differences of opinion among the educated medical specialists. This should be viewed as a sort of academic exercise, which is wonderful fun for those of us who would like to think we are experts in the disorder, but I think nonetheless that industry can get on with the job on the information which is already available.[16]

Asbestosis

Mr. Hutchinson [president of the Asbestos Workers Union] and I were talking this morning of one of the men in his union, who, before he died, used to walk backward. I had never seen this before until I began to care for asbestos workers.

You may wonder why asbestos workers walk backwards. They don't always walk backwards. It is only going upstairs. They are so short of breath that after two steps they have to sit down. It is easier to go up a flight of stairs backwards than walking up. It is a terrible way to die.

—*Testimony of Dr. Irving J. Selikoff, before the Senate Subcommittee on Labor, May 5, 1970*[17]

Asbestos is a soft and highly flexible material that is virtually indestructible: it is resistant to chemical action and temperature changes, and is fireproof. No wonder, then, that asbestos is ubiquitous in the modern industrial environment. The cement industry consumes vast quantities of asbestos. It is used in roofing, insulation board, and pipes. Asbestos insulation often lines the air ducts of air-conditioning systems. The material is also found in safety cloth, papers, felts, millboard, floor tile, plastics, many paints and sealants, and in automobile brake linings and clutch facings. Even some kinds of talcum powder contain a form of asbestos. There is no factory or home without asbestos products, and there are few lungs that are not exposed to microscopic asbestos fibers.

The lung disease asbestosis was first described fifty years ago, although the ancient Greeks were aware of its ill effects on health. Dr. W. E. Cooke described a severe case of lung scarring (pulmonary fibrosis) in a thirty-three-year-old woman who had worked in an English asbestos-textile factory from the time she was thirteen.[18] The woman's autopsy showed extensive scarring and dense strands of fibrous tissue connecting the lungs and the pleural membranes surrounding them. Cooke also noted "curious bodies" in the areas of fibrosis, and these quickly came to be known as "asbestos bodies."

Dr. Irving J. Selikoff of the Mt. Sinai Environmental Sciences Laboratory has conducted the most definitive studies of asbestos workers to date. In 1962 he tested 1,117 asbestos workers from two local unions and found radiological evidence of asbestosis in half of them.[19] What makes this discovery so alarming is the fact that asbestosis is irreversible.

He also began research on 632 asbestos-insulation workers who had been members of two local unions in 1942. As of June 30, 1971, 425 were dead.[20] Mortality tables for the general population indicated that only 285 deaths were to be expected. In addition, 47 of these deaths were specifically due to asbestosis.

In the past, because asbestosis victims lived at the very brink of pulmonary incapacity, they would often die of a simple cold, bronchitis, or pneumonia. Modern antibiotics have reduced the severity of these complications and tend to postpone immediate death. Unfortunately, one major result of this postponement is that asbestos workers, instead of dying of asbestosis, now live long enough to develop asbestos-induced cancer.

A causal relation between asbestos and lung cancer was first proposed in 1935 by Dr. Kenneth Lynch, professor of pathology at the Medical College of South Carolina, who described the case of an asbestos weaver who had died with what were then two very unusual diseases: asbestosis, and lung cancer.

Dr. Selikoff has also found a definite relationship between asbestos exposure and lung cancer. Using standard mortality tables, Dr. Selikoff noted before the Senate Subcommittee on Labor that lung cancer was found seven times more often in asbestos workers than in the general population.

There is more to this grim story. A cancer that had previously been so rare that is was not coded separately in the World Health Organization's International Classification of Diseases was discovered to be the cause of death in an alarming number of cases in South Africa in 1956 and 1957. The cancer, mesothelioma, can be induced by lower levels of asbestos exposure than will cause asbestosis. It is a cancer of the pleural linings of the lungs and chest, or of the peritoneum—a similar membrane that lines the abdominal cavity. Mesothelioma is diffuse and covers the entire lung; it cannot be surgically or medically removed. In his testimony before the subcommittee Dr. Selikoff noted that there has not been a single cure of mesothelioma in the world. In the data on the 632 men from 1942 to 1967, 20 deaths were due to mesothelioma compared to 1 out of 10,000 in the general population.[21]

Asbestos-insulation workers are not the only victims of mesothelioma. For example, a woman born in a town near asbestos mines in South Africa died at the age of fifty-seven from pleural mesothelioma, even though she moved away from the region at the age of five.[22] But during those early years she attended a kindergarten near an asbestos dump and used to slide down the dump on her way home from school (as did her playmates, two of whom also died of mesothelioma). Dr. Muriel L. Newhouse and Dr. Hilda Thompson, of the Department of Occupational Health at the London School of Hygiene and Tropical Medicine, found that of seventy-six confirmed deaths of mesothelioma, thirty-one had worked with asbestos, eleven had lived within half a mile of an asbestos factory, and nine others were relatives of asbestos workers. Seven of these nine were women, and most of them washed their husbands' work clothes regularly, one of them brushing her husband down every evening when he came home from work "white with asbestos."[23]

More recently, Dr. Selikoff and his associates have linked asbestos exposure to cancer of the stomach, colon, and rectum. As reported in *The New York Times* of October 5, 1972, continuing observation of the 632 insulation workers has revealed 41 deaths from gastrointestinal cancer, or three times the expected number under normal circumstances.

At present there are some 36,000 asbestos-insulation workers in this country.[24] The accumulated data of the mortality experiences of the 632 men Dr. Selikoff has followed from 1943 through June 30, 1971, indicate that 11 percent of them have died of asbestosis,[25] and that nearly 38 percent of them died of cancer.[26] One in five deaths has been due to lung cancer, one in ten to gastro-intestinal cancer, and mesothelioma has accounted for nearly one in ten.[27]

Yet asbestosis and lung cancer take twenty to thirty years to develop. What we are seeing now is the result of exposure thirty years ago, and during the last thirty years the amount of asbestos in use has greatly increased. In the most recent report on asbestos-worker mortality rates, Selikoff and his associates said:

> The occurrence of two deaths of mesothelioma among the men who began work after 1943 is an unhappy omen. Until these, it was possible to hope that the experience of the men who had begun work before World War II would not necessarily be repeated among their colleagues who first began work after the war.[28]

In late September, 1972, Dr. Selikoff called particular attention to the possibility that men and women who worked in shipyards during World War II might also now be developing mesothelioma. The British have already begun to notice mesothelioma cases among former shipyard workers who had been indirectly exposed to asbestos particles, and Dr. Selikoff has found several cases involving American shipyard workers.

The American Conference of Governmental Industrial Hygienists, a private group that sets safe limits for industrial exposures over a forty-hour work week (called Threshold Limit Values, or TLV's), determined as of 1969 that the limit for asbestos should have been twelve fibers per cubic centimeter of air (roughly a thimbleful). Each fiber is more than five microns (five one-thousandths of a millimeter) in length.

When asked by a Study Group member what he thought of the TLV Dr. Selikoff said, "It's marvelous, marvelous for people with nothing to do. They can fondle it and touch it. The problem is that it's *never* measured." Dr. Selikoff said that in the men he has studied, who have worked 4.5 million working hours, there have been only seven dust counts. Selikoff said the TLV does have significance for engineering design: the smaller the number chosen, the more expensive it is to meet the limit.

"I have never seen a safe level," Dr. Selikoff said. "I'm not interested. No one does dust counts. Safe working conditions have to be engineered."[29] For example, instead of mixing asbestos cement in the open, it can be done in plastic bags, with *no* dust.

Dr. Selikoff estimates that if one includes asbestos-textile workers and people engaged in manufacturing other asbestos products, there are probably more than one hundred thousand workers who are regularly and directly exposed.[30] Moreover, according to Dr. William H. Stewart, the former Surgeon General of the United States, the asbestos-insulation workers are sharing their exposure to some extent with more than 3.5 million other construction workers, such as steam fitters, electricians, welders, carpenters, and plumbers.[31] Even office workers are exposed to asbestos-lined air-conditioning ducts.[32]

And they are sharing their exposure with the rest of the urban population, as the demolition and construction of buildings often send clouds of asbestos particles down upon an unwary public.

Other industrial carcinogens

The fate of [chimney sweeps] seems singularly hard; in their early infancy they are most frequently treated with great brutality, and almost starved with cold and hunger; they are thrust up narrow and sometimes hot chimneys, where they are bruised, burned, and almost suffocated, and when they get to puberty become peculiarly liable to a most noisome, painful, and fatal disease. . . .

It is a disease which always makes its first attack on, and its first appearance in, the inferior part of the scrotum, where it produces a superficial, painful, ragged, ill-looking sore, with hard rising edges; the trade calls it soot-wart. I never saw it under the age of puberty, which is, I suppose, one reason why it is generally taken both by patient and surgeon for venereal.

—Percival Pott, first describing occupational cancer of the skin in 1775[33]

Asbestos is not the only industrial material causing cancer of the lung and of other sites. Coal-tar products, containing aromatic hydrocarbons, have long been recognized as inducing cancer of the skin, lungs, and bladder. Among workmen involved in the process of carbonization, such as producing gas for household and industrial use, and in the production of coke, essential to steelmaking, there have been reports of unusual cancer experience.

While doing a long-term mortality study of over fifty thousand steelworkers, Dr. J. William Lloyd and his associates found that the mortality rate from respiratory cancer for men employed in the coke plants was twice the rate generally observed among steelworkers.[34] They further found that all of the differences in mortality rates for this disease were due to a three-fold excess for black workers and were somewhat puzzled by "the apparent differential in mortality for white and nonwhite workers." To better understand why men in the coke plants had twice as much lung cancer as most steelworkers, and why blacks had three times the amount, they divided the coke plant population into those who worked around the coke ovens and those who did not. While 7.5 deaths from lung cancer were expected among the general steelworker population, there were 20 deaths at the coke ovens. Others working in the coke plant, but not at the ovens, had about the same mortality rate as the general steelworker population. Further, those who worked at the coke ovens for five or more years showed total mortality 17 percent in excess of expectation and a lung cancer mortality three and one-half times greater than that predicted (27 versus 7.6). Those who worked five or more years on top of the ovens, rather than on the side, showed a ten-fold risk of lung cancer. Fifteen lung cancer deaths were observed among the 132 men in this group, compared to 1.5 expected.

Dr. Lloyd and his associates found that, in their base year, 89 percent of nonwhite workers were employed at the coke ovens compared to only 32 percent of white workers. Furthermore, they found a higher turnover rate for whites than nonwhites, resulting in greater long-term coke oven exposure for blacks.

The authors concluded that:

> The apparent differential in respiratory cancer rates for white and nonwhite coke plant workers, reported in the fourth paper in this series, is accounted for by differing distributions by work area and the unusually high lung cancer risk for topside workers.[35]

Airborne radiation is a significant cause of death among uranium miners in the United States. In the next twenty years an estimated 6,000 uranium miners will die as a result of radiation exposure, primarily from lung cancer.[36] A mortality study of a group of 3,414 white underground uranium miners published in 1969 found that there were 398 deaths observed, versus 251 deaths expected, on the basis of rates for the white male population of the area.[37] Violent deaths (120 versus 50.5 expected) and respiratory cancer (62 versus 10 expected) largely accounted for this excess of 147 fatalities. There was no excess of cancer deaths observed in the first five years after the start of mining, and most deaths occurred after ten years of exposure. This study underscored the relationship between the amount of exposure and the rate of cancer, which had been previously discussed in the medical literature, and while it was found that uranium miners who smoked cigarettes had an excess of lung cancer ten times greater than nonsmoking miners, smoking alone did not explain the marked excess of lung cancer.

A high incidence of cancer of the lungs has been reported in men employed in refining nickel. From 1948 to 1956 the rate of nickel workers dying from lung cancer was 5 times that of the rest of the population (and from cancer of the nose, 150 times).[38] A survey of the causes of death in the chromate-producing industry in the United States found that death from cancer of the lung occurred among chromate workers about 25 times more frequently than the expected rate.[39] A relationship between bladder cancer and the manufacture of certain dyes has also been established.[40]

Cancer of the skin, particularly of the face, neck, limbs, groin, and navel, afflicts those who work with coal products such as tar, pitch, asphalt, anthracene (used in the manufacture of synthetic dyes and in wood preservation), and creosote (used in wood preservation).

Mineral oils made from petroleum are also a particular hazard. Used in a wide variety of engineering processes, they are ubiquitous throughout modern industry—serving as lubricants in machine tooling to dissipate heat, reduce friction, and wash away metal chips. There are over 750,000 gallons of industrial lubricating mineral oils consumed in the United States every year.[41]

Most discussion of mineral oils appears in the foreign medical literature. According to the November 7, 1970, issue of *The Lancet*, mineral oil has been responsible for two epidemics of scrotal cancer in England. The first, which began in the 1880's, ravaged workers in the cotton industry; it affected men who leaned across spindles lubricated by shale oil containing the carcinogen benzpyrene. The second broke out in the late 1940's or early 1950's among men who worked with automatic lathes that were both cooled and lubricated by circulating mineral oil.[42] There is little mention of scrotal cancer in the American literature, either because of an extremely low incidence in this country, or because it has not yet been recognized here.

One recent discussion in an American journal, *Industrial Medicine and Surgery*, evidences the cavalier attitude toward industrial diseases often found in this country. The article acknowledges that cases of scrotal cancer may still occur among workers exposed to mineral oils other than shale oil. After stressing the importance of personal hygiene, the author suggests that the most important thing workers can do is examine themselves and report anything unusual, instead of having routine twice-yearly examinations. He says that such examinations would be too expensive. And as for substituting safe synthetic oils for carcinogenic ones, he says:

Although an ideal to be aimed at, this cannot at present be regarded as a practicable procedure. The carcinogens in mineral oil have not as yet been precisely identified, and it is therefore not possible to say categorically that certain oils are "safe" whilst others are "carcinogenic."[43]

Under certain circumstances the rapid movement of machinery has the effect of turning mineral oil into an aerosol mist, the inhalation of which may cause lung cancer. An article in *The Lancet* noted the coincidence of three cases of men who had survived carcinoma of the scrotum attributed to exposure to mineral oil only to succumb later to cancer of the lung. The authors observed:

> Without further information, particularly about the working habits of the men concerned, it is unwarranted to assume that the respiratory and digestive cancers resulted from exposure to oil mist in the engineering shops where the men's skins were sufficiently exposed to give them scrotal cancer; but this must be regarded as a strong possibility.[44]

Poisoning from metals

The doctors cannot diagnose my disease. I am afraid it is cadmium poisoning. It is running through my whole body. Pain eats away at me. I feel I want to tear out my stomach. Tear out all of my insides and cast them away.

A lathe operator in a Japanese zinc company recorded these words in her diary shortly before she threw herself from a moving train. The authorities listed her as a suicide victim. Two years later, an autopsy confirmed the fear expressed in her diary and revealed that she had been suffering from cadmium poisoning.[45]

In California, an experienced welder suffered an overexposure to cadmium fumes and went home ill. He developed a severe cough and chest pains, which he attributed to "welder's fever." Fewer than four days later, he was found dead in his bedroom. The autopsy showed an extreme hemorrhagic congestion of the lungs, and the cause of death was determined to be cadmium poisoning.[46]

Cadmium is a metal used as a protective coating on iron,

steel, and copper. The firing or welding of these substances releases cadmium fumes, which have no pronounced odor or immediate effect, and are so toxic that they were once considered for use in chemical warfare. Inhalation of these fumes causes a severe and unusual form of emphysema, and damage to the kidneys and bone marrow. There have been relatively few deaths from acute cadmium poisoning officially reported in the United States, but it is quite conceivable that cadmium fatalities may go unrecognized as such.

Mercury is a silvery white metal, liquid at ordinary temperatures. The ancient Romans were aware of its poisonous properties. The Mad Hatter in *Alice in Wonderland*, like so many hatters of his day, was suffering from classical mercury poisoning, contracted from the mercury used in the process of stiffening felt. The Hatter exhibited the signs of what is technically known as mercurial erethism: hallucinations, delusions, and mania. Erethism is the most characteristic symptom of early chronic mercury poisoning.

Mercurial tremor usually begins in the fingers, but the eyelids, lips, and tongue are affected early. It then progresses to the arms and legs. In the United States it has been known as "the Danbury shakes" and more generally as "the hatters' shakes."

The earlier and more ambiguous symptoms are salivation and tenderness of the gums and mouth. Recent exposure to mercury can often be determined by analysis of urine for mercury content. But this method will not tell how much mercury a person has been exposed to, especially over a long period of time.

Mercury is generally absorbed in the system through the respiratory tract in the form of vapor, fume, dust, or mist, or through skin contact. Mercury combines with the sulfhydryl groups to inhibit the action of essential enzymes and concentrates—primarily in the kidneys and liver, and also in the brain.

Although hatters are not particularly exposed to mercury these days, it is widely used in the manufacture of thermometers and gauges and is ubiquitous in chemical laboratories and hospitals across the nation.

The first citation issued by the government under the Occupational Safety and Health Act of 1970 was to the Moundsville, West Virginia, chlorine plant of Allied Chemical Corp. for excess mercury exposure. Concentrations of mercury vapor more than eight times the Threshold Limit Value were found in the plant.[47] Dr. Sidney M. Wolfe, serving as consultant to the Oil, Chemical and Atomic Workers Union (OCAW), found that urine levels of mercury in workers were as much as six times above the safe limit.[48] Dr. Wolfe said that in a trip to Moundsville he found what he called "the classic neurological symptoms" of mercury poisoning among many of the thirty-four employees he interviewed.[49] Stewart Udall and Jeff Stansbury, in their syndicated column, said:

> As early as 1965 the men in Allied's mercury cell room knew something was wrong. Preventive maintenance in the eleven-year-old plant was slipping.
> The workers could smell chlorine—which meant the concentration of the gas in the air was at least three times higher than the national safety limit. Mercury vapor was leaking from gaskets. Liquid mercury was dripping to the floor and vaporizing at room temperatures.
> One by one, the men began showing those subtle signs of chronic mercury poisoning that only a specialist can distinguish from everyday mental disstress: a bad temper, shyness, fatigue, depression. In more advanced stages mercury poisoning causes tremors, brain damage, and death. Robert L. Fisher, a mercury cell worker, now hospitalized, has the tremors.
> Allied cooled the workers' fears in the late 1960's by insisting that the plant was clean and that metallic mercury was safe. It gave the men more medical tests. The results of the tests were not disclosed.[50]

During this period of time Allied removed the men from the cell room when the mercury levels in their urine rose above the safety threshold and then sent them back when the levels dropped again, ignoring the cumulative buildup of mercury in the body.

Another metal associated with widespread harmful occupational exposure is lead. The industrial uses of lead include the manufacture of paints, batteries, pipes, bullets, chemicals, and tin cans, the soldering of automobile radiators, and the casting of type in various processes such as linotyping and electrotyping. A common procedure in many types of chemical plants—lining tanks and reaction vessels with metallic lead and repairing these linings from time to time—is a source of significant exposure to lead fumes, especially when the work has to be done on the inside of extensively corroded receptacles. Also, men who enter gasoline storage tanks (where leaded gasoline has been stored) to clean or repair them risk a highly perilous exposure to tetra-ethyl lead and its decomposition products.

The effects of lead, like other metal poisons, are usually slow until an adequate amount accumulates. Most of the lead is stored in the bones and is harmless while there. The lead which is either not stored, or is released, however, can damage the blood, brain, nerves, and possibly the kidneys. Symptoms vary from abdominal pain due to colic, to fatigue, weakness, and paleness. Some people have partial loss of function of their hands and suffer headaches and an impairment of vision. Children poisoned by lead frequently incur mental retardation.

In 1969, at the National Lead Co.'s Hoyt plant in Granite City, Illinois, approximately 100 of 145 employees were contaminated with high lead levels in their blood. Eighty employees were taking medication, and 10 previously had taken medication for excessive lead levels. Three employees were then currently hospitalized, and at least 7 had been hospitalized in the past several years. A total of 70 percent of the employees had suffered contamination from lead to the extent of requiring medication, and two had retired because of severe lead poisoning.

One of the hospitalized employees described his condition as follows: "My joints hurt, I can hardly talk anymore—I shake all the time—sometimes I don't think I can get up in the morning." This employee, who was also suffering from severe stomach pains, had been taking medication for the past fifteen years for excessive lead in his blood. Yet he was severely "leaded," totally and permanently disabled. His nervous system was destroyed, and his mental processes were deteriorating rapidly. For example, his speech was severely impaired.[51]

The National Lead Co. denied knowledge of any history of lead poisoning among its employees and claimed its medical records showed no lead poisoning for the employee just mentioned.[52]

Gases

> The masks don't keep all this out of your system. When I
> go home at night, I have roast beef and the fumes. I can
> taste it, still taste it when I get home. If I have dessert, I
> can still taste some of the chemicals. It's serious, these
> chemicals.
>
> —Harold Smith, Local 8–447, Oil,
> Chemical and Atomic Workers,
> OCAW Conference, March 29, 1969[53]

Although the first citation under the Occupational Safety and
Health Act confined itself to mercury metal poisoning, the men
in the same plant also complained of excess chlorine. Steve
Wodka, an OCAW union spokesman, said:

> From what we can tell, the company, since the citation, is
> making efforts to clean up the plant. But we feel that the
> Secretary of Labor acted capriciously in not even mentioning
> the chlorine problem at this plant. The fact that this citation
> concerns itself only with mercury poisoning and does not
> even mention chlorine is just incredible.[54]

The Oil, Chemical and Atomic Workers Union has had a
longstanding problem with chlorine, one of the stronger of a class
of elements known as the halogens, which includes fluorine, bromine, and iodine. Chlorine is often an ingredient in the manufacture of chemicals, solvents, pesticides and herbicides, plastics
and fibers, refrigerants and propellants, pulp and paper, and is
used in water and sewage treatment, and in textile bleaching. It
has a Threshold Limit Value of 1 part per million (ppm) and
cannot be smelled until it is 3.5 ppm. So, if chlorine can be
smelled, it is at least three or four times too strong to be safe.
This level of exposure causes irritation of the eyes, nose, and
throat, and sometimes headache from irritation to the nasal
sinuses. In high concentrations chlorine can be fatal. It was used
as a poison gas in World War I, and about 10 percent of the
survivors of the gas attacks later developed emphysema.[55] A
book published by the American Conference of Governmental
Industrial Hygienists, *Documentation of Threshold Limit Values*,
says that "a ceiling of 1 ppm is recommended to minimize

chronic changes in the lungs [emphysema], accelerated aging, and erosion of the teeth."[56]

In the National Lead Co.'s plant in Sayreville, New Jersey, the odor of chlorine was marked.[57] Dan Staley was overcome by chlorine gas in July, 1967, while he was sitting outside the conversion building after lunch. He was taken to Perth Amboy Hospital for medical treatment, lost two days' work, and required three months of medical treatment outside the plant. Robert Hawkins was gassed at least ten times in 1968. When asked what he was doing he replied, "Pressure checking. Sometimes just being in an area where a leak happened and just not getting out in time." He was treated with cough medicine and said he usually felt sick for a day or more, depending on the amount he had inhaled.

T. F. Wisneski was gassed during a pipe-fitting job and, when asked if he required medical treatment, said, "No. Just sick to my stomach and chest pains."

Tom Callaher was overcome and given cough syrup. He said there were extensive gas fumes in his area and described what happened when he attempted to climb off a platform in the area:

> I stumbled and hit my head on a rail. Fumes were so bad that other operators could not get to me to help me. I managed to get down, trying to hold my breath, and proceeded to cough and throw up in the street. The foreman went inside and brought out some cough medicine. After sitting in the street awhile, I went inside and threw up some more. After a few hours I felt a little better.

Chlorine is what is known as an irritant gas. Other irritant gases include ozone, sulfur dioxide, ammonia, nitrogen dioxide, phosgene, fluorine, and hydrofluoric acid—the strongest acid known to man.

Ozone, a form of oxygen, is a colorless or pale blue gas with a pungent odor, and it can be deadly. It endangers welders using inert-gas, shielded-arc welding devices. The discharge of high-voltage electrical equipment also produces this gas. Exposure to ozone may also occur from ozone generators used in industrial processes to disinfect or control the growth of fungus, molds, and bacteria in food processing, or to purify water. Ozone is also a key component in oxidant smog derived from hydrocarbons, sunlight, and nitrogen dioxide. A Public Health Service pamphlet on ozone says:

> There is no exposure to ozone without some risk to health. On this basis, all unnecessary exposure of humans to any concentration, however small, should be avoided. The danger of undesirable health effects far outweighs any benefits presumed to be derived from the industrial or institutional use of ozone for control of odors or bacteria in air.
>
> Even in low concentrations, inhaled ozone may cause dryness of the mouth, throat irritation, coughing, headaches, and pressure or pain in the chest, followed by difficulty in breathing. Varying in individuals, and depending upon the concentration of ozone and the period of exposure, ozone can produce many other injuries. It impairs the sense of smell, disguises other odors with a continuous odor of ozone, alters taste sensation, and reduces the ability to think. Ozone also depresses the nervous system, thus slowing the heart and respiration and producing drowsiness and sleep.
>
> Exposure to relatively small concentrations for an hour can cause serious cough and fatigue. Exposure to sufficient quantities for two hours results in a marked reduction in the capacity of the lungs. Exposure for four hours can cause the lungs to fill with fluid and start bleeding. . . .[58]

The long-range effects of ozone exposure show that ozone can significantly alter the structure of cells and tissues, most importantly in the lungs. These changes include a breakdown of cell membranes and the production of lung fluid (edema). Rabbits exposed for one hour a week to doses of ozone that were narrowly within the tolerable range showed a progressive breakdown of alveolar walls, leading to emphysema. Although the levels were not high enough to produce acute effects, insidious chronic change still took place. Besides producing pulmonary problems, ozone also causes premature aging and accelerates the growth of lung tumors.[59]

Sulfur dioxide and smoke are the main constituents of smog. Sulfur dioxide concentration of 0.5 ppm killed more than a hundred New Yorkers in 1967.[60] The highest concentration of sulfur dioxide recorded during the "Killer Smog" of 1952, which took the lives of four thousand Londoners, reached 1.3 ppm. The Environmental Protection Administration warns that as little as 0.1 ppm causes adverse health effects. A recent article on the subject has noted that:

> Based on acute episodes of high pollution, such as those which have occurred in the Meuse Valley, Belgium, Donora, Pennsylvania, London, New York, etc., there is convincing evidence that high levels of smoke and sulfur dioxide, and other pollutants lead to excessive mortality and morbidity in populations, specifically for respiratory and cardiac diseases, affecting those forty-five years or older especially.[61]

The author adds, "There is a good deal of evidence that sustained lower levels of pollution affect health adversely,"[62] and that significant correlations have been found between levels of smoke and sulfur dioxide and mortality and morbidity from all causes, as well as from certain respiratory diseases.

If levels of sulfur dioxide produced by factories damage the health of urban residents, one may wonder what happens to the workers inside these same plants. A study of 681 workplaces, located in New Jersey, Pennsylvania, and Detroit, indicated that, "The air which workers breathe on their jobs is foul and dangerous—much worse than the air outside the plants, and polluted air is bad enough on the outside."[63] In addition to finding excess silica dust, particulate matter (such as asbestos and foundry dust), oil mist, sulfuric acid, iron and zinc oxides, coal tar, and lead, the study reported that the average in-plant sample of sulfur dioxide was 5.6 ppm, more than four times the highest level in London's "Killer Smog." The annual average measure producing bad health effects on city residents is 0.04 ppm. (The Threshold Limit Value for sulfur dioxide is 5 ppm, or fifty times higher than the Environmental Protection Administration warning level, ample proof of how inadequately workers are protected against harmful job exposures.)

Some gases, such as carbon monoxide, hydrogen sulfide, and hydrogen cyanide, exert a specific chemical action on either the blood or body tissues and can cause brain damage or death by depriving the body of oxygen. Although hydrogen sulfide and hydrogen cyanide have somewhat more dramatic effects than carbon monoxide (inhalation of a high concentration of any of them is fatal), exposure to carbon monoxide is a much more widespread occupational hazard.

For example, in early 1968 in the Sayreville plant of the National Lead Co. (the same plant with excess chlorine), a new process developed for making titanium dioxide involved the use of carbon monoxide. Within the first six months of operation, excess exposure to the carbon monoxide took its toll. Two men collapsed on the job, two other men suffered permanent brain damage (one of them is totally disabled now), and one man was killed. All five were twenty-five years old or younger.[64]

Carbon monoxide is particularly pernicious because it simply cannot be detected by the human senses. It poisons by combining with the hemoglobin in the blood and interrupting the normal oxygen supply to the body tissues and brain. When severe and prolonged deprivation combines with unconsciousness, irreversible degenerative changes in the brain may occur. These changes may include paralysis, amnesia, and difficulty in speech, coordination, and comprehension.

The chronic effects of prolonged exposure to carbon monoxide are no less important: headache, nausea, dizziness, sensory deprivation, lack of coordination, and general debility. The permanent effects of prolonged carbon monoxide exposure are less clear than the effects of acute exposure, but there is evidence that prolonged exposure produces severe strain on the heart. First, the decrease in oxygen in the blood makes the heart pump faster to increase the amount of blood going through the lungs to receive oxygen. Second, long-range exposure produces more hemoglobin and red blood cells to balance the reduced oxygen-carrying capacity. This increases the density of the blood and makes it harder to pump.[65]

The present Threshold Limit Value for an eight-hour exposure to carbon monoxide is 50 ppm. No monitors or alarms had been triggered at National Lead in the exposed part of the plant until after the death from monoxide. Then they were discovered to be set at 100, 200, and one was even set at 400 ppm —eight times the recommended level.[66]

Even at levels of 50, 40, and 30 ppm, subtle physiological changes can be measured in humans exposed to carbon monoxide. For instance, the ability both to see a dim spot of light against a wall and to add a column of numbers decreases.[67] Reaction time is slower, and this has very serious implications for the workplace. During a delicate or complex operation requiring rapid judgment, carbon monoxide may impair ability to perform adequately and may lead to accidents.

Solvents

> *Unbeknownst to me, the [toluol] fumes in the tank were getting to me, although I could smell the odor of the toluol at the time. I was in such a hurry to get the tank clean and all I needed was a couple of more swashes around, and I would be done, but those couple of swashes caused me a fright. I was really frightened because what happened was, I just passed out. I just passed out completely, just fell asleep on the floor, and later I was taken out and walked around the block like a dog to clear my system of the fumes I had inhaled.*
>
> *—Tom Grasso, Local 8–584, Oil,*
> *Chemical and Atomic Workers,*
> *OCAW Conference, March 29, 1969*[68]

Solvents are chemicals that dissolve other chemicals. They also can dissolve the fats and oils in the skin and thereby cause various kinds of skin disease (dermatitis). In 1968 the U. S. Surgeon General estimated that 800,000 people suffer from some form of occupationally related dermatitis each year.[69] A second harm caused by solvents occurs as a result of their tendency to vaporize. Some solvent vapors can be quite toxic to the lungs, respiratory tract, and other organs of the body.

For example, benzene is a powerful, fast-drying solvent used in printing. In 1939, a scientific study found that three rotogravure printing plants in New York City, where 350 men were employed, used about fifty thousand gallons of benzene per month.[70] The benzene content of the ink solvents varied from 10 to 75 percent by volume; the benzene content of the thinners varied from 20 to 80 percent by volume. The benzene concentrations in the workroom atmosphere ranged from 11 to 1,060 ppm.

(The present Threshold Limit Value of benzene is 10 ppm.) A total of 332 workers were examined, and of these 130 showed varying degrees of benzene poisoning. Among a group of 102 cases completely studied, 22 exhibited benzene poisoning of a severe degree (6 requiring hospitalization), and 43 were early cases. The men were exposed to benzene through evaporation from open troughs on the machines, from spills that resulted in pools on the floor, and from the paper itself. Benzene was also used to clean the cylinders and troughs, and the workers used it to clean themselves at the end of the day.

William Martin, assistant director of the Special Project on Respiratory Diseases, International Printing Pressmen and Assistants Union of North America, told one Study Group member that the *Washington Post* discontinued using benzene only three or four years ago.[71]

At present 9.3 billion pounds of benzene are produced every year in this country.[72] In addition to the printing industry, the chemical, drug, rubber, boot and shoe, paint and varnish, and dry cleaning industries use benzene as a solvent. It is also a starting and intermediate material in the synthesis of numerous chemicals and is a constituent of motor fuels.

When benzene comes in contact with the skin (and lungs and nose), it tends to dissolve the skin's fats because of its dehydrating and defatting action. When benzene is inhaled in high concentration, it will accumulate in the brain and other fat-containing tissues and has a narcotic effect on the nervous system. Inhalation may cause exhilaration, followed by drowsiness, dizziness, nausea, and headache. If there is loss of consciousness, deep respiratory anesthesia and paralysis will follow. If the victim is not moved away from the exposure, death will be certain.

When benzene is inhaled in small amounts over a prolonged period of time, it will cause liver injury, and the blood-forming functions of the bone marrow will be destroyed, leading to debilitation, need for constant transfusions, vulnerability to infection, and death from aplastic anemia. Furthermore, there is some evidence that benzene causes leukemia.[73]

Benzene is the simplest of the aromatic hydrocarbons (one

constituent of coal tar products and mineral oil). It is closely related to other solvents such as toluene and the xylenes, which are used as solvents for synthetic rubber, paint and lacquers, and as motor fuel constituents. Toluene is more acutely toxic than benzene because it has stronger narcotic effects on the nervous system.[74] It is not more toxic chronically, however.

Toluene and xylene also tend to dissolve fats from the skin, lungs, and nose, often resulting in the development of a rash. The present Threshold Limit Value for toluene is 200 ppm. *The Documentation of Threshold Limit Values* says toluene has moderate effects. "Fatigue, moderate insomnia, and restlessness resulted from continuous exposure for eight hours to 200 ppm." At 100 ppm it has slight effect, according to this standard source, and at 50 ppm it has minimal effect. Yet in the Soviet Union, the Threshold Limit Value for toluene is 25 ppm,[75] another example of stricter occupational exposure limits often set by other countries.

Because benzene is so toxic it is often replaced with other chemicals. For printers, many of the replacements were formulated with methanol, or wood alcohol.[76] Unfortunately, methanol damages the optic nerve and has a noxious effect on the kidneys, liver, heart, and other organs. Handling methanol has caused many cases of blindness. It is also highly inflammable.

Many nonflammable fluids contain carbon tetrachloride (popularly known as "carbon tet"), which is widely used in printing, as a chemical intermediate, as a fumigant, as a solvent for metal degreasing, as a suppressant for mixtures of more inflammable materials, and as a dry cleaning solvent. Inhalation of high concentrations of carbon tet causes depression of the central nervous system. Dizziness, headache, depression, confusion, and loss of consciousness may follow. More important, however, are the long-term chronic effects: severe injury of the internal organs, particularly of the kidney and liver.[77] The Food and Drug Administration has banned the sale of carbon tetrachloride for use in the household.[78]

In some people carbon tet is extremely toxic, since even small concentrations will cause severe kidney damage resulting in destruction of the kidneys. For example, a woman in Albuquerque, New Mexico, while cleaning her purse outside her home during a windy day, inhaled a small amount of carbon tet.[79] This caused her kidneys to cease to function for several weeks. She survived, since her kidneys were able to survive the assault. Not every carbon tet victim is this fortunate.

Although carbon tetrachloride is the most deadly of the chlorinated hydrocarbons, many others, often found in cleaning fluids, are potentially dangerous. These include trichloroethylene, perchlorethylene, chloroform, trichlorethane, and ethylene dichloride.

Another class of organic solvents is the ketones, which are widely used as solvents for dyes, oils, fats, tars, waxes, and many natural and synthetic resins and gums. The ketones are irritating to the skin because of their defatting action. Ketone vapors are narcotic; initial irritation of the eyes, nose, and throat may be followed by drowsiness, loss of consciousness, and, with sufficient depression of the respiratory center, death. Chronic exposure leads to lung irritation characterized by emphysema, limited tissue congestion usually centering in the liver, kidneys, and sometimes the brain. Irritation of the intestinal tract has also been recorded.[80]

Epoxy compounds

Epoxy compounds are important ingredients in the manufacture of plasticizers, special solvents, surface-active agents, synthetic resins, cements and adhesives, and fine chemicals. The most commercially important of the epoxy compounds are the epoxy resins, first synthesized in the 1930's. They are versatile, tough, highly adhesive, have a low degree of shrinkage, and an indefinite shelf-life. Many can be worked with at room temperature. The resins are widely used as surface coatings and adhesives, especially for metals, and in the treatment of textiles to make them more crease resistant.

Unfortunately, the resins are extremely irritating, and contact with fumes of the epoxy resins has caused dermatitis of the face, eyelids, and neck. Inhalation of vapors or aerosols can lead to acute pulmonary edema, or fluid in the lungs. The reported incidence of dermatitis among those having prolonged or unusual contact runs between 10 to 60 percent. Sensitization is not uncommon and may run as high as 2 percent of the exposed population.

Some epoxy compounds cause central nervous system depression. Repeated exposures to ethylene oxide, one of the epoxy compounds that has widespread commercial use, can paralyze the lower extremities. There have also been reports of blood cell changes as well as indications that epoxy compounds might be carcinogens.[81]

Toluene–2, 4–Diisocyanate, or TDI, is a component of polyurethane foam, which is commonly utilized in household appliances, airplane construction, mattress and upholstery padding, soundproof walls, linings for overcoats and sleeping bags, insulation against heat loss, life preservers, fish-net floats, packaging, and a number of other products. Since the introduction of TDI in the manufacture of synthetic foams, doctors have reported a number of cases of severe pulmonary effects among workers, usually beginning after a latent period with repeated exposure.

The current Threshold Limit Value is .02 ppm, a ceiling *not* to be exceeded. However, an eighteen-month study of American factory workers recently found that acute changes in lung capacity occurred in workers exposed to TDI at levels *less* than the TLV.[82] The study also determined that these acute changes were not reversed overnight and documented cumulative changes in lung capacity in workers during the eighteen months of the study. The authors said that whether these changes would be reversed with a decrease or cessation of exposure is unknown.

Pesticides

REP. DANIELS: *Why did these people who work on the farm become sick?*

MRS. OLIVERAS: *Because the day before, they spray the field. The next morning they put the people in the field and they took about six of these workers to the doctor and the doctor talked to the people. He said they get sunstroke. The people are going to get sunstroke by seven o'clock in the morning?*

> —*Testimony of Lupe Oliveras, member, United Farm Workers Organizing Committee, before the House Select Subcommittee on Labor, November 21, 1969*[83]

Pesticides present an extremely serious problem to farm workers and their families, as well as to the general population. In California agricultural workers suffer the highest occupational disease rate in the state,[84] and pesticide poisoning is certainly one of the most serious hazards they encounter.

One extremely dangerous class of pesticides includes organophosphates such as azodrin, demeton, ethion, Guthion, parathion, methyl parathion, phosdrin, TEPP (tetraethyl pyrophosphate), and Thimet. The Germans first developed organophosphates during World War II for use as nerve gas, and the present pesticides are closely related to the German compounds.

The organophosphates kill by inhibiting cholinesterase, a critical enzyme in nerve and muscle cells. Normally, impulses pass from nerve to nerve, or from nerve to muscle, with the help of the chemical mediator, acetylcholine, which is then destroyed by cholinesterase. If the cholinesterase is inhibited, the acetylcholine continues to act, and the body's nervous system becomes hyperactive. At first the symptoms are similar to the flu: nausea, vomiting, muscle cramps, sweating, and headaches. If the dose is large enough to completely inhibit cholinesterase, tremors, spasms, convulsions, and death will result. It takes a very small dose to kill a human being.

Farm workers face exposure to organophosphates in a number of ways. One common hazard is accidental exposure while spraying crops:

> A young sprayer was found dead in the field in the tractor which had been pulling his spray-rig. He had been pouring and mixing parathion concentrate into the spray-rig tank. Parathion has an estimated fatal dose of about nine drops orally and thirty-two drops dermally. In the process of mixing

the concentrate, the worker contaminated his gloves inside and out. He rested his gloved hands on his trousers as he pulled the rig to apply the spray. Parathion was absorbed through the skin of his hand and thighs. He began to vomit, an early symptom of parathion poisoning. He could not remove his respirator and he aspirated the vomitus. The diagnosis of poisoning was confirmed by postmortem cholinesterase tests.[85]

Another common exposure results from picking crops with pesticide residues still on them:

In 1963, ninety-four California peach harvesters were poisoned by parathion residues on the foliage of the orchards in which they worked. By law, there is a waiting period between pesticide application and crop harvesting so that poison will have deteriorated to the point where residues on the food are within safe limits. Apparently, in this case, the law was obeyed. Nevertheless, the workers got sick. Medical tests indicated that the probable cause of the illness was a poison in the spray residue which is a product of parathion but is even more toxic than parathion itself.[86]

Children are also the victims of pesticide poisoning. In fact, between 1951 and 1965, approximately 60 percent of the deaths from pesticide poisoning in California were among children:

A three-year-old Mexican-American girl and her four-year-old brother were playing around an unattended spray-rig next to their mother who was picking berries on a large ranch. The four-year-old apparently took the cap off a gallon of 40 percent TEPP pesticide left on the rig. One drop of pure TEPP swallowed or on the skin will kill an adult—this child weighed thirty pounds. The little girl put her finger in the bottle, then in her mouth. She became unconscious and was dead on arrival at the hospital where her mother rushed with her. The rig operator was apparently not told to remove the pesticide, much less the rig, immediately after its use; the children's mother was not told of the danger of the pesticide; and there was no supervised safe place to leave the children while the mother worked.[87]

Herbicides, or weed-killers, also present a dangerous occupational exposure. 2,4,5–T (or 2,4,5–trichlorophenoxyacetic acid) is the most notorious of the herbicides, and with good reason. 2,4,5–T has long been used as a defoliant in Vietnam. The American Association for the Advancement of Science (AAAS) has asserted that one-fifth to one-half of South Vietnam's mangrove forests have been "utterly destroyed," and that perhaps half the trees in the mature hardwood forests north and west of Saigon are dead.[88]

Furthermore, there is a distinct possibility that 2,4,5–T causes birth defects in human beings. In a report to the National Cancer Institute in 1966, the Bionetics Laboratory revealed that 2,4,5–T causes a high rate of fetal mortalities and birth defects in laboratory mice. The results were summarized three years later:

> Tested more extensively than other pesticides, 2,4,5–T was clearly teratogenic [caused birth defects] as evidenced by statistically increased proportions of abnormal features . . . in particular, cleft palate and cystic kidneys were significantly more prevalent. . . . The incidence of fetuses with kidney anomalies was threefold that of the controls, even with the smallest dosage tested.[89]

Yet the use of 2,4,5–T in Vietnam reached its peak in 1967, a year after the first Bionetics Laboratory results were completed.[90]

Further evidence unfortunately confirms that 2,4,5–T causes birth defects in human beings. Besides noting the environmental damage to Vietnam from 2,4,5–T, the AAAS found an alarming increase in stillbirths and deformities in newborn children coincident with large-scale spraying. In Tay Ninh, a very heavily defoliated province, stillbirths in 1968 and 1969 were at a rate of 64 per thousand, compared with a countrywide average of 31.2 per thousand. Saigon Children's Hospital experienced a disproportionate rise in birth defects in 1967 and 1968.[91]

The U. S. Government failed to make public the effects of 2,4,5–T. The 1966 results of the Bionetics Laboratory tests were suppressed for three years. Only after some digging by Anita Johnson, a member of the Nader Study Group on the Food and Drug Administration, and subsequent complaints and inquiries by a number of scientists, was the issue brought into the open in October, 1969, when the White House science adviser, Lee A. DuBridge, announced that 2,4,5–T usage would be restricted after January 1, 1970.[92] DuBridge's announcement also mentioned that the Defense Department had been directed to restrict its use of 2,4,5–T to "areas remote from population." The Defense Department said no change would be needed in its policy because its present policy conformed to this directive.[93]

The ban did not go into effect on January 1, 1970, however, and the test results of the Bionetics Laboratory were disputed by Dow Chemical, manufacturers of 2,4,5–T. The results were then replicated by the National Institute of Environmental Health Sciences (NIEHS)[94] and, as a result, in April, 1970, 2,4,5–T was suspended for use in Vietnam "pending a more thorough evaluation of the situation."[95]

Also as a result of the NIEHS findings, the federal government banned liquid forms of the herbicide for use around the home within the United States and suspended all uses of all formulations in lakes, ponds, and ditch banks. Nonliquid forms could still be used, and the ban did not affect the use of 2,4,5–T for control of weeds and brush on range, forest, pasture, and other nonagricultural land. At present, 90 percent of 2,4,5–T sales remain unaffected.[96]

The foregoing account of contemporary health hazards covers only a fraction of the actual damage industry is inflicting upon American workers. Of the estimated half-million substances found in the occupational environment, only 450 have become subject to TLV's, and many of these maximum exposure limits are sadly out of date.[97]

It is also necessary to keep in mind that in addition to noxious substances, hazardous stresses are taking an unreasonably high toll. For example, the number of American workers experiencing noise conditions that may damage their hearing is estimated to be in excess of 6 million, and may even reach 16 million.[98] Some 50 percent of the machines in use in heavy industry generate noise levels that are potentially deleterious to hearing.[99]

We have already emphasized the critical importance of worker awareness of the dimensions of the occupational health and safety problem. But a grasp of these dangers to his physical well-being will not be sufficient to protect him. Though many particular health hazards are new and result from modern technology, the desirability of securing a reasonably safe and health-ful work environment has been recognized for many decades. Efforts to achieve this goal in the United States began in the nineteenth century.

Therefore, a full understanding of why these efforts have not succeeded is of paramount significance. For not to heed the past is to risk, if not assure, its repetition.

3

A History of
Human Sacrifice

Throughout the formative years of industrial development in the United States, American workers and their families made a vital contribution to the prosperity and growth of big business —a private subsidy coined in blood, broken bones, broken health, physical pain, mental anguish, and the ultimate trauma of death.

At the turn of the twentieth century, the human toll exacted by industry began to provoke a great public outcry that soon stirred the national conscience. Work-accident statistics told a grim story. As one writer put it in 1912:

> As many men are killed each fortnight in the ordinary course of work as went down with the "Titanic." This single spectacular catastrophe appalled the civilized world and compelled government action in two hemispheres; while the ceaseless, day-by-day destruction of the industrial juggernaut excites so little attention that few states take the trouble to record the deaths and injuries.[1]

The year 1907 was particularly tragic. There were 3,242 fatalities in the anthracite and bituminous coal mines. On December 6, 361 men were killed in a mine explosion in Mononghah, West Virginia. Thirteen days later, a mine explosion in Jacobs Creek, Pennsylvania, claimed 239 lives. In 1907, some 4,534 railroad workers were killed on the job.

Crystal Eastman's book, *Work-Accidents and the Law*, published in 1910, became a best seller and helped fuel popular indignation. Especially effective was the frontispiece, a calendar dating from July, 1906, to June, 1907. Under each day was a check-mark for each industrial death occurring that day in Allegheny County, Pennsylvania, the heartland of the American steel industry. The relentless consistency of the daily toll, and the final total of 526, conveyed a gruesome message.

Statistics told only part of the story. Knowledge about industrial diseases, for example, was scant and often rudimentary. ("Printers die fast and die young" was a conclusion of one early study of health problems of members of the International Typographical Union of North America.[2]) Also, it was a matter of common knowledge that a substantial number of work injuries and deaths were unreported. Who could forget this passage from Upton Sinclair's classic, *The Jungle*, a work of fiction, yet totally consonant with the facts of the day:

> It was said by the boss at Durham's that he had gotten his week's money and left there. That might not be true, of course, for sometimes they would say that when a man had been killed, it was the easiest way out for all concerned. When, for instance, a man had fallen into one of the rendering tanks and had been made into pure leaf lard and peerless fertilizer, there was no use letting the fact out and making his family unhappy.[3]

The first decade of the twentieth century was the golden age of muckraking in American magazines, and a talented corps of writers went to work describing in the most vivid terms the unsafe, unhealthy conditions of the nation's factories and workshops. They particularly deplored the financial hardships suffered by disabled workers and their families. In the words of one of the best of these journalists:

> For the agony of the crushed arm, for the torment of the scorched body, for the delirium of terror in the fall through endless hollow squares of steel beams down to the death-delaying construction planks of the rising skyscraper, for the thirst in the night in the hospital, for the sinking qualms of the march to the operating table, for the perpetual ghostly consciousness of the missing limb—for these things and for

the whole hideous host of things like them, following upon
the half-million accidents that happen to American working
men every year, there can be no compensation.[4]

The reason injured workers had to deplete their own meager
savings and even turn to public relief to defray the costs of accidents was that the legal system virtually insulated employers
from liability for work injuries and deaths.

During the growth era of American industry, corporate
leaders felt that "progress" justified certain inevitable costs, and
they viewed occupational accidents as inevitable costs of progress.
Yet historical determinism is a highly superficial explanation of
the human toll. It smacks of testimony before a Congressional
investigating committee in 1912 by a U. S. Steel executive, who
answered complaints about the excessive hours employees in the
steel industry were forced to work with the pious cant that hours
were set "by the laws of nature."[5]

There was nothing inevitable about the formulation and
execution of the rules of law relating to industrial accidents.
Legislatures and courts faithfully reflected the interests of those
who dominated the economy, and who were adamantly unwilling to have the cost of work injuries interfere with profits and
growth. They held to the proposition that property rights took
precedence over human rights, an ethic that permeated the entire
social fabric in the nineteenth century.

To sanction this policy of shielding industry from the economic burden of industrial mishaps, the courts applied legal
principles that made it virtually impossible for injured workers
to shift their losses to the enterprises that employed them. A
lawsuit for money damages required that the employee prove
that he was injured because of the fault of his employer. In
cases involving serious accidents, it was often impossible for an
injured employee to determine what happened to him, let alone
whose fault it was.

In addition to his opportunity to deny that he was in any
way at fault, the employer could raise any of three legal
defenses, which, if proven, would completely exonerate him—
even if the accident had been caused by his negligence. He could
argue that the accident had resulted, in whole or in part, from

the employee's own carelessness; that the employee had voluntarily chosen to be exposed to the work hazard that caused the accident; or that a fellow employee was responsible for the accident. None of these three legal defenses made much sense or was in any way just.

In situations where the employee's carelessness was very slight—momentary inadvertence, for example—in contrast to the employer's maintenance of unsafe machinery or working conditions, it was manifestly unfair to deny the employee any recovery at all, especially if his injuries were serious. In addition, as one writer noted in 1910 in describing several categories of worker carelessness in the steel industry:

> . . . human powers of attention, universally limited, are in [the worker's] case further limited by the conditions under which work is done—long hours, heat, noise, intense speed. For the reckless ones we maintain that natural inclination is in their case encouraged and inevitably increased by an occupation involving constant risk; recklessness is part of the trade.[6]

In applying the employer's second defense, the so-called doctrine of assumption of the risk, the courts took the position that by remaining on the job in the face of known dangers, the worker had bargained away his right to recover damages from the employer. Inherent in this doctrine was the idea that the worker, if he did not wish to encounter the risk, could always quit his job and move to another. In the light of the harsh realities of economic conditions during this era, the ubiquitousness of dangerous machinery, and the uniformity of hazardous work practices in every industry, this reasoning was clearly untenable.

The so-called fellow-servant rule was the least defensible of the three defenses. To bar recovery to an employee injured by the carelessness of a co-worker made no sense at all, except as a blatant device to protect employers from legal liability.

Though damage suits usually took the form of trials by jury, the presiding judge could take any case away from the jurors and direct a verdict for the employer if he found as a matter of "law" that the employer was not negligent, or that one of the three defenses was applicable. In this way the trial judge could eliminate the possibility that jurors might sympathize with the injured worker and bend legal rules in order to do justice.

A final factor militating against the worker with a valid claim was the law's delay. It might take two years or more before his lawsuit was scheduled for trial. This placed upon him a serious financial burden, which attorneys for employers often exploited by postponing and prolonging the trial as long as possible. As a result the worker was generally forced because of economic hardship to settle his claim for a fraction of what it was worth. Since he also had to pay his own lawyer out of any settlement, his final recovery was seldom adequate.

Various estimates have been made, ranging from 70 to 94 percent, of the accident and death claims that the law left uncompensated. From 1906 to 1912, the U. S. Steel Corporation is said to have lost no more than six verdicts.

> During the year ending July, 1907, Jones and Laughlin paid nothing to the families of seven, funeral expenses for ten, and over a thousand dollars to only two families of its toll of twenty-five dead. In Allegheny County 88 out of 158 injured married men received no compensation, and 23 of 33 men permanently disabled were given less than one hundred dollars.[7]

From the employer's point of view, therefore, the cost of industrial accidents was in no way onerous. Most companies purchased liability insurance, which included the services of insurance company lawyers to defend against claims. Approximately one-third of their premium dollars went to pay claims by injured workers, with the remainder eaten up by the insurance company's expenses and profits. The big industrial concerns were self-insurers, which meant that they employed their own lawyers and paid out their own liability claims. With such little economic incentive for safety, it is easy to see how employers came to consider industrial accidents as "inevitable."

Another important causative factor helping to keep the work environment both unsafe and unhealthy was the prevalent philosophy of "rugged individualism." In the early days of industrial development in the United States, private initiative served as the handmaiden of progress. The "Horatio Alger" ethic held that any individual could achieve financial prosperity if he worked hard enough at it. Personal responsibility for success implied personal responsibility for failure, from which it followed, albeit tenuously, that industrial accidents were failures for which the worker should be held responsible. The legal principles governing work accidents reflected this attitude, which had taken root in the days of the small workshop. The courts were unwilling to modify these principles in the face of the increasing complexities of the manufacturing process, which put working conditions beyond the control of the individual worker. This was especially true in the case of industrial exposures to dust, gases, chemicals, and other noxious substances and stresses, which destroyed the health of workers who were unaware of what was happening to them.

"Rugged individualism" did not make much of a contribution to safety in the steel industry because of the practice of appointing to in-plant supervisory positions men who had come up through the ranks on their own initiative and merit.[8] These supervisors had personally run all the risks of injury and death inherent in the industry and, somehow having survived, had become hardened to them. Their stomachs did not turn at the sight of mangled or charred bodies. For them the task at hand was to keep production up and costs down. They could not be expected to promote safety in the plants.

The wide exploitation of immigrant labor also contributed to the precedence of property rights over human rights in the matter of work safety. The prevalent attitude toward immigrants was brilliantly captured in a 1911 muckraking magazine article describing the author's visit to the office of a construction engineer, which overlooked a recently excavated railroad cut:

> "To think," I exclaimed, "that not a man was killed."
> "Who told you that?" asked the young assistant.
> "Why, it's here in this report sent to the newspapers by your press agent. He makes a point of it."
> The young assistant smiled. "Well, yes, I guess that's right," he replied. "There wasn't anyone killed except just wops."
> "Except what?"
> "Wops. Don't you know what wops are? Dagos, niggers, and Hungarians—the fellows that did the work. They don't know anything, and they don't count."[9]

Dr. Alice Hamilton, a pioneer in the field of industrial medicine, confirmed this attitude in her autobiography:

> Yesterday I visited doctors and druggists and hospitals [in Colorado during the 1910's]. I am amazed to see how lightly lead poisoning is taken here. One would think I was inquiring about mosquito bites. When I asked an apothecary about lead poisoning in the neighborhood of the smelter, he said he had never known a case. I explained that that was incredible, and he said: "Oh, maybe you are thinking of the Wops and Hunkies. I guess there's plenty among them. I thought you meant white men."[10]

Foreign workers arriving in great waves of immigration at the turn of the century were quickly converted into industrial cannon fodder. Coming from peasant societies, and unable to speak English, they greased the wheels of industry with their blood.

> The accident rate for non-English-speaking employees at the South Works steel mill from 1906 to 1910 was twice the average of the rest of the labor force. Almost one-quarter of the recent immigrants in the works each year—3,273 in the five years—were injured or killed. In one year 217 Eastern Europeans died in the steel mills of Allegheny County.[11]

Because of language problems and unfamiliarity with their legal rights, foreign-born accident victims and their families found it uncommonly difficult to press claims against their employers.

The muckrakers focused most of their fire on the financial plight of the injured worker and tended to slight the problem of industrial accident prevention. This may have been due in part to the prevailing belief that work mishaps were inevitable, and in part to the assumption that making employers pay for the costs of these mishaps would sufficiently motivate them to reduce the frequency and severity of work injuries.

Some of the states did enact safety laws, dating from the passage in 1877 of a Massachusetts statute requiring the guarding of hazardous parts of machinery, such as shafts and gears. But effective enforcement was something else. A 1910 survey of the situation in Pennsylvania concluded that state safety inspectors were too few and were politically appointed; they were not very zealous about their work and failed even to keep public records of prosecutions of safety code violations.[12]

On the industrial health front, "In 1910 it was possible to record only the appointment of the Illinois Occupational Disease Commission, the completion of an investigation of phosphorous poisoning in the match industry, and the holding of the First National Conference on Industrial Diseases. . . ."[13]

The federal government was slow to take any steps to help safeguard American workers. In 1906 many European countries had signed a treaty calling for the prohibition of the manufacture, importation, or sale of matches made from white phosphorus. It was not until December 6, 1910, that President Taft asked Congress for legislation to prevent the "frightful diseases" contracted by workers engaged in the manufacture of these matches.* But it took two more years for Congress to get around to placing a prohibitive tax on the domestic sale of white phosphorus matches and to forbid their import or export. In 1910, in the wake of countless human disasters in the coal mines, Congress created a Federal Bureau of Mines, but merely gave it a mandate to conduct investigations without compulsory powers. The only safety problems in which Congress took any sustained interest were those arising from railroading, and beginning in 1893 a number of railroad safety laws were passed.

Prior to 1911, various efforts were made to shift the costs of industrial accidents from individual workers and their families. Many states passed employer liability laws that modified the common-law legal rules and made it easier for injured employees to recover money damages from their employers. Groups of workers pooled their scant resources to establish funds from which they or their families might draw modest payments in the event of disability or death. Some of the larger corporations furnished medical and financial benefits to injured employees without regard for fault. The U. S. Steel Corp., sensitive to the winds of impending reform, began an extensive safety program in 1908 and, two years later, introduced its own compensation plan for industrial accident victims.

* White phosphorus causes necrosis of the jaw. "It was the most distressing of all occupational diseases because it was very painful and was accompanied by a fetid discharge which made its victim unendurable to others."[14]

But these piecemeal measures failed even to approximate a solution. Several states appointed commissions to study the problem. They paid specific attention to reforms that had been initiated by Bismarck in Germany in 1884, and in England in 1897. The latter became the model for a proposed new approach to industrial accidents in the United States—workmen's compensation.

A statement attributed to England's Lloyd George nicely encapsulates the underlying philosophy of workmen's compensation: "The cost of the product should bear the blood of the workman." Under the new system the employer would be liable regardless of fault for work injuries and deaths "arising out of and in the course of the employment," but would pay to the employee or his family cash benefits limited by statute, as opposed to the full compensatory damages recoverable under the common law. This formula involved a *quid pro quo*: the employer gave up his three legal defenses, while the worker surrendered the possibility of obtaining full damages. The cost of work accidents would pass to the employer, who could then absorb them by adjusting the price of his product. In addition, claims under workmen's compensation were to be processed by an administrative agency, which would grant prompt recoveries to workers whose injuries were covered by the act. In theory, this was the way the new statutes were supposed to operate.

Congress passed a workmen's compensation statute for federal employees in 1908. Two years later New York enacted the first state workmen's compensation act, which the New York Court of Appeals held unconstitutional. But several other states passed their own compensation acts in 1911, and these, as well as a subsequent, revised New York compensation statute, overcame constitutional objections and were upheld by the courts. By 1920, all but eight of the states had passed workmen's compensation laws.

The new statutes did not cover every worker or every occupation. Nor did they provide recovery for industrial diseases. But they did succeed in shifting from employee to employer some part of the total economic burden of industrial accidents and thereby reduced to a more "tolerable" level a problem that had previously reached critical dimensions.

Though accident frequency rates did show a marked decrease after the enactment of the new legislation, it has not been possible to document any clear relationship between workmen's compensation and work safety. Prof. Walter F. Dodd, reviewing the situation in 1936, concluded that, "The operating forces which, since 1911, may have produced increase or decrease of industrial injuries are not capable of measurement; but it is perhaps safe to say that workmen's compensation has had little effect upon the result."[15]

Two points are worth noting. First, the subsiding of accident rates was at least in part the inevitable result of the institution of the most rudimentary and inexpensive safety practices, which had up to then been generally ignored by employers; and some of these reforms, especially in the areas of elevator and boiler safety, came about as a result of state safety laws and regulations. Second, most of the compensation acts ignored occupational diseases, and the accident frequency rates cited as evidence of progress took no account of silent, subtle assaults on the worker's health. Despite all this, in the years immediately following the passage of the new compensation statutes, there were many who resorted to *post hoc, ergo propter hoc* reasoning and hailed the salutary effect of the new laws upon the work environment.

The emergence of workmen's compensation did not mark any revolutionary change in values, nor did it signal the end of the ascendancy of property rights over human rights. Though the humanitarian aspects of the new statutes were often proclaimed, the truth of the matter was that workmen's compensation made excellent business sense, and for that reason powerful business interests backed its adoption. Dissatisfaction with existing methods of work-accident compensation had reached such a level that some reform was inevitable. Businessmen feared the liberalization of employer's liability laws, which would have limited the employer's common-law defenses and made it much easier for injured workers to recover full compensatory damages. Workmen's compensation would be much less costly to industry, particularly because state legislatures would be relatively easy to manipulate, and thus compensation benefits could be kept at a low level. The preexisting system also presented a political issue around which workers might be mobilized. This was a time when labor unions were beginning to grow in strength and militancy. Although organized labor assigned a low priority to accident compensation and prevention, far-sighted business leaders saw the need to eliminate sources of conflict between labor and management and to deprive the emerging labor movement of a potentially explosive issue.

Industry, moreover, had begun to understand the true costs of easily avoidable work accidents. When corporate leaders took into account property damage to machinery, the disruption of the work process after an accident, and the training of replacements for skilled workers disabled or killed on the job, they came to realize the economic waste caused by these mishaps and initiated their own safety movements. They were motivated by self-interest, as expressed by Judge Elbert H. Gary, chief executive of U. S. Steel, who once promised his board of directors, "If you will back us up in it [a proposed safety drive], we'll make it pay."[16] These safety movements thus derived from dollars-and-cents judgments reflecting a more rational application of the same property-oriented philosophy that produced the deplorable conditions in factories and workshops at the turn of the century.

At the same time, as reformist pressures for government regulation of the workplace increased, businessmen took the position that such regulation should remain within the jurisdiction of the states. It was no problem for corporate interests to see to it that state legislatures kept workmen's compensation benefits at low levels, lagging far behind rising prices and wages and scarcely adequate to meet the most basic economic needs of the employee and his family. The administrative agencies that were supposed to process claims in a speedy, efficient manner found themselves bogged down in disputes—as employers contested claims on a myriad of technical grounds. The abundance of high-powered legal talent at the service of industry and the absence of countervailing representation for injured workers helped to produce state court decisions that restricted the new compensation statutes.

The first compensation statutes made no mention of occupational disease. The underlying reason for this omission was a widespread conviction that disease was more properly a subject for health insurance. Some reformers thought that since disease might affect the public generally, the problem should be handled by a system of public health insurance, a view that seemed to ignore the causal relationship between industrial processes and occupational diseases. When efforts to pass health insurance laws failed, the workmen's compensation system became the last hope for employees disabled by job-related diseases. Some of the compensation statutes were amended to provide benefits for specifically listed diseases only. Others were changed to provide general coverage of all occupational diseases. Still others were modified to include statutory definitions of the occupational diseases for which compensation would be awarded.

Gaps in industrial-disease coverage naturally affected the attitudes of employers toward disease prevention. As one writer observed:

> The automobile companies "do not worry," as one safety director of a certain General Motors plant put it to the writer, about industrial diseases or the steady weakening of the worker's health arising out of months or years of work in the factory, and the consequent exposure to poisons that slowly break down health. These diseases are "noncompensable" in the state of Michigan and hence do not entail payments by the employers, as do accidents.[17]

Accident and disease prevention fared little better than compensation. Though some state agencies did good, enduring work in the fields of boiler and elevator safety, the general picture was discouraging. As one critic writing in the 1930's noted:

> Characteristic of these first laws regarding health and safety, as of all the early excursions into the field of social legislation, was a great inadequacy. Too often they were enacted with only the most flagrant abuses in mind and too often, also, the interested organization and interested individuals have allowed their zeal to evaporate as soon as the laws were placed on the statute books. The matter of enforcement, so these trusting souls seemed to feel, would more or less take care of itself.[18]

The same writer also noted that:

> . . . these laws have been largely specific, passed to meet the requirements of specific situations; whereas for a century

past industry itself has been undergoing one change after another in incredibly swift succession. No wonder the laws have been out of date almost before they were passed.[19]

A hard look at the various laws designed to promote job safety and the compensation of workers disabled at work would have uncovered numerous shortcomings. Nonetheless, the problems of occupational safety and health languished in the limbo of neglect for more than half a century. The muckraking era in American journalism, essentially the work of intellectual reformers, failed to generate a grass-roots constituency and passed with the advent of World War I. The assumption prevailed that the new workmen's compensation laws were causing accident rates to decline. Nobody paid heed to the inadequacy of existing state health and safety laws and their nonenforcement. Nor was there any attempt to arouse among the workers themselves an awareness of the full dimensions of the health and safety issue, nor a sense of responsibility for their own self-protection and survival. The labor movement was gradually building political muscle, but it did not press for the reduction of the frequency and severity of work accidents and diseases.

The Great Depression further decelerated interest in occupational safety and health. As the economic crisis worsened and unemployment mounted, workers became willing to face any kind of physical hazard in order to hang onto their jobs. This served only to diminish the already feeble pressures on industry to eliminate or reduce such hazards. Conditions were ripe for a disaster of the first magnitude, and it happened at Gauley Bridge, West Virginia, in 1930 and 1931.

In retrospect, it is incredible that the story of the digging of the tunnel near Gauley Bridge did not break until 1935. Although much controversy was to surround the calculation of the project's human cost, a U. S. Public Health Service official testifying before a Congressional committee in 1961 put it at 476 dead and 1,500 disabled.[20] Yet it took five years from the time construction began for nationwide attention to focus on the tragedy, and the full facts did not emerge until a year later in the course of a Congressional hearing.[21]

The New Kanawha Power Co., a subsidiary of Union Carbide, secured a permit from the West Virginia legislature to build a hydroelectric plant in the southern part of the state, ostensibly for the purpose of supplying badly needed energy to the impoverished surroundings. In fact, the primary function of the plant was to power the nearby Electro-Metallurgical Co., another Union Carbide subsidiary engaged in the electro-processing of steel. Water from the New and Kanawha rivers was to be diverted through a tunnel to be built through a mountain near the town of Gauley Bridge. In 1929, the Rhinehart-Dennis Co. was chosen as the subcontractor on the tunnel construction project.

The original plan called for a tunnel width of thirty-two feet. But when the rock formation at the tunnel site was found to contain a high silica content, New Kanawha Power changed the specifications to allow a forty-six-foot width, so that silica excavated from the tunnel might be shipped to the Electro-Metallurgical Co. to be used in the processing of steel.

Actual construction began in 1930. Rhinehart-Dennis recruited its labor force from Pennsylvania, Georgia, North and South Carolina, Florida, Kentucky, Alabama, and Ohio, as well as from the surrounding hills of West Virginia. Most of the workers were black and unskilled. The pay scale began at fifty cents an hour, but as the Depression wore on, it dropped to as low as thirty cents an hour. Blacks paid fifty cents a week for their living quarters, ten-by-twelve-foot shacks, each of which housed twenty-five to thirty men, women, and children. They paid seventy-five cents a week for medical care, while whites paid fifty cents a week for the same service.

Working conditions strained credulity. Gasoline-powered trains filled the tunnel with carbon monoxide, poisoning the workers and making them drowsy. The dust was often so thick that workers couldn't see ten feet in front, even with the headlight of a train. (Indeed, occasionally trains collided with each other.) Though West Virginia mining law required a thirty-minute wait after blasting, workers were herded back into the tunnel immediately after a blast. The foremen at times had to beat them with pick handles to get them to return. The silica content of the rock being blasted was extremely high. Though New Kanawha Power warned its engineers to use masks when entering the tunnel, no one ever told the workers to take precautions.

Increasing numbers of workers became progressively shorter of breath and then dropped dead. Rhinehart-Dennis contracted with a local undertaker to bury the blacks in a field at fifty-five dollars per corpse. Three hours was the standard elapsed time between death in the tunnel and burial. In this way, the company avoided the formalities of an autopsy and death certificate. It was estimated that 169 blacks ended up in the field, 2 or 3 to a hole.

Whenever the doctors did perform an autopsy, they concluded that the cause of death was tuberculosis, pneumonia, or "tunnelitis." But the actual precipitating cause of most of the deaths was silicosis, which transformed lungs into a mass of scar tissues as a result of the inhalation of dust with a high silica content. Before work on the tunnel began, it was a known fact that prolonged exposure to dust with a 25–30 percent silica content could cause silicosis. The silica content of the dust in the tunnel was 90–95 percent. There was little ventilation. Men were dying from nine to eighteen months after brief exposures to the silica. Toward the end of the project, some workers purchased their own respirators for $2.50. The purchasing agent for Rhinehart-Dennis was overheard to say to a respirator salesman, "I wouldn't give $2.50 for all the niggers on the job."[22] The paymaster was also overheard to tell the superintendent, "I knew they was going to kill those niggers within five years, but I didn't know they was going to kill them so quick."[23]

The deadly dust, however, was not color-conscious. The foreman who herded the men into the tunnel died. The assistant superintendent died. White workers died. Fifty percent of the tunnel workers died or were in the various stages of silicosis. Gauley Bridge and surrounding towns became villages of the living dead. In July, 1934, an official of the Federal Emergency Relief Administration visited the black community in one of the towns. Of the 43 males in the community of 101, 33 were dying of terminal silicosis. The official recommended relocation. "It is unadvisable socially," he noted, "to keep a community of dying persons intact."[24]

At the time of the Gauley Bridge disaster, the West Virginia workmen's compensation act did not cover silicosis, so that those workers who wanted to press claims had to sue at common law. Under West Virginia law, as interpreted by that state's Supreme Court, lawsuits against Rhinehart-Dennis and New Kanawha Power had to be filed within one year of the harmful exposure to silica dust, a patently unfair rule because many of the workers did not become aware until much later that they had contracted the lung disease. Some two hundred cases were thrown out of court because of the one-year rule.

The trials themselves were a macabre burlesque. A Rhinehart-Dennis foreman testified in a wheezing voice that there had been no dust in the tunnel. (He later died of silicosis.) The chief of West Virginia's Mines Department testified that he had observed no dust in the tunnel in 1930 or 1931, yet in 1931 he had written letters urging the company to do something about the dust condition. One of the jurors was held in contempt for riding home every evening in a company car. Rumors of jury tampering abounded, and workers found themselves unable to win the necessary unanimous verdicts. The juries consistently voted eleven-to-one or ten-to-two in favor of the workers, which meant that a new trial would be necessary. Finally, 167 of the suits were settled out of court for $130,000, with one-half going to the workers' attorneys. Payments were meager, with blacks receiving $80 to $250, and whites from $250 to $1,000. One law firm representing workers accepted a $20,000 payment on the side from the companies in order to make the lawyers more amenable to a settlement. When this bit of chicanery came to light, the lawyers were compelled to pay back half of this sum, which was then distributed among their clients.

In January of 1936, Vito Marcantonio, a freshman Republican Congressman from New York City, convinced a subcommittee of the House Committee on Labor to hold hearings on the Gauley Bridge disaster, and the grim facts unfolded. The subcommittee concluded that Rhinehart-Dennis had been negligent, in that ". . . the whole driving of the tunnel was begun, continued, and completed with grave and inhuman disregard of all considerations for the health, lives, and future of employees."[25] West Virginia Congressman Jennings Randolph (now a Senator) filed a minority report disassociating himself from the majority's conclusions, lamely lamenting that ". . . the lack of safety equipment contributes in a marked degree to the silicosis problem."[26]

On the other hand, Senator Rush Dew Holt of West Virginia showed no sympathy for the corporate interests responsible for the Gauley Bridge affair. In his testimony before the subcommittee he called it: "the most barbaric example of industrial construction that ever happened in this world. That company well knew what it was going to do to these men. The company openly said that if they killed off those men there were plenty of other men to be had."[27]

Despite the public outcry over the slaughter of the tunnel workers, the fact remains that nothing much happened as a result of the hearings. The subcommittee in its report recommended that Congress fund a study of silicosis, but this request never made its way out of the Committee on Labor. Secretary of Labor Frances Perkins called a series of "Washington Town Meetings," at which industry and labor talked about the silicosis problem. Some of the states began to amend their compensation acts to make silicosis a compensable occupational disease. Workmen's compensation insurance rates in these states skyrocketed, and both industry and labor urged a reduction in benefits to head off the removal of certain plants to other states whose laws did not compensate workers who contracted the disease. The West Virginia compensation statute was amended to include silicosis several months before the Congressional hearings began, too late to help any of the casualties of the Gauley Bridge affair. (Under the amendment to the West Virginia act, workers could collect benefits for silicosis only if they had been working for two years or more on the job that exposed them to the silica dust. The legislators thereby made certain that in the event of a repeat of a Gauley Bridge disaster, the workers could still not collect any compensation!) No punitive action was ever taken against Rhinehart-Dennis, New Kanawha Power, or Union Carbide.

The failure of the horror at Gauley Bridge to precipitate any effective, broad-based action against the scourge of work accidents and diseases was a clear indication of the low status of the job safety and health issue during the Depression years. The advent of World War II brought a revival of interest in the problem, as the federal government sought to reduce occupational accidents that might adversely affect the war effort.

But two attempts to enact federal legislation on the subject came to naught. In 1940 a bill to provide federal grants-in-aid to states that set up industrial hygiene units in their labor departments failed in the Senate Committee on Labor and Education.[28] And in 1943 a bill that would also have provided federal funds to state agencies administering labor laws for the purpose of improving safety and health conditions in industry suffered a similar fate in the House Committee on Labor.[29] Opponents of both bills raised the specter of federal control over job safety as an unacceptable encroachment upon states' rights. Bureaucratic jealousies also intruded as public health officials and others opposed putting health functions in state labor departments.

In the post-war era, American industry modernized and expanded, developing new products and processes at a dizzying clip. Workers were exposed to chemicals and chemical compounds whose effects, especially in the long run, were unknown. Manufacturing methods created novel stresses that ceaselessly battered employees. With technology totally untethered, corporate officials made no place on their balance sheets for expenditures for research that could accurately project what impact these innovations would have on the working man. Job safety and health laws inadequate to deal with traditional employment hazards could hardly have been expected to protect employees from the new dangers.

There were several more futile attempts at federal legislation. The most interesting was Senator Hubert H. Humphrey's Accident Prevention Act of 1951, which would have set up a bureau in the Department of Labor to develop safety standards, and an independent board to enforce them. Another Senate bill introduced at the same time would have channeled federal funds to state labor departments to help them promote job safety.[30] Neither bill reached the Senate floor. Finally, in 1962 the General Subcommittee on Labor of the House Committee on Education and Labor held hearings on two bills designed to provide grants to the states for job safety programs.[31] The committee reported out an Occupational Safety Act, which was then bottled up and blocked in the House Rules Committee. It is interesting to note that none of these post-war bills paid any heed to occupational disease, an indication of a pervasive unawareness of the problem.

Indeed, general ignorance, fortified by the silence of the few who fully realized what was happening, was an underlying reason for the absence of progress in the struggle against work accidents and illnesses. But this excuse certainly did not apply to workmen's compensation. Studies and statistics were readily available to tell anyone with eyes to see that the state compensation systems, after more than half a century of existence, were badly shortchanging the workers.

For example, the Bureau of Labor Standards of the U. S. Department of Labor has been periodically publishing a list of standards recommended for incorporation into state compensation statutes by groups such as the American Medical Association, the Council of State Governments, and the Department of Labor. These suggested standards are quite modest, as might be expected from their sponsors. They include compensation for agricultural workers, full coverage for occupational diseases, and full medical benefits for work injuries and diseases. In 1961, a comparison between nineteen model provisions and the compensation acts of fifty states, the District of Columbia, and Puerto Rico revealed that of a total of 988 categories (nineteen standards multiplied by fifty-two jurisdictions) 58 percent failed to meet the recommendations.[32] In its 1967 publication, the Bureau of Labor Standards used sixteen model provisions. The fifty-two jurisdictions were still 58 percent deficient.[33] If one removes from the 1961 list the three standards not included in the 1967 list, the 1961 rating rises to a deficiency of 61 percent, which means that measured by the sixteen standards used in 1967, the fifty-two jurisdictions made a 3 percent improvement in six years—not exactly an encouraging performance. Indeed, one cannot overemphasize the stark fact that in a span of nearly six decades, the workmen's compensation system in the United States has progressed to the point of being 58 percent deficient.

Cash benefits under workmen's compensation are supposed to replace a percentage of the injured employee's wage loss. In actual operation, this percentage has turned out to be a mere trifle. In 1962, in Illinois, workmen's compensation was found to pay for only 18 percent of the injured worker's estimated present and future wage loss.[34] Using 1956 figures, a 1961 study concluded that workmen's compensation benefits under the California statute replaced a mere 12.2 percent of the median net loss suffered by the survivors of the victims of fatal industrial accidents in that state.[35] On a national level, the study found that thirty-three states replaced 20 percent or less of the estimated loss.

In August of 1968, a Presidential task force submitted to the Council of Economic Advisors a confidential report entitled "The Workmen's Disability Income System—Recommendation for Federal Action," a document that both the Johnson and Nixon administrations have suppressed. Professor Monroe Berkowitz, the task force chairman, and Dr. John F. Burton, Jr., used an updating of some of the materials in the report as the basis for an article in the October, 1970, issue of the *Industrial and Labor Relations Review*. Curiously, they made no reference to the task force or its work.

The report disclosed that in forty-three states the maximum allowable 1968 workmen's compensation benefits for permanent total disability did not meet the 1966 poverty standard-of-living level as stated in the *Social Security Bulletin*. The 1970 article reported that in 1968, thirty-eight states paid out maximum benefits for permanent total disability that were less than the 1968 poverty-level figure.[36]

According to the Berkowitz-Burton article, "The data suggest that far less than one-half of the workers are receiving the benefit levels the workmen's compensation program is alleged to provide, namely, about two-thirds of previous wages."[37] In New Jersey, for example, a worker in 1940 could have recovered a maximum weekly benefit of $20 for temporary total disability. The average weekly wage in the state at that time was $29.45. In 1966, he would have received $45 a week for temporary total disability, as compared with an average state weekly wage of $123.81. In 1968, New Jersey raised its statutory maximum to $86, which was 63 percent of the state's average weekly wage. But forty other states in 1968 had statutory maximums that provided 60 percent or less of the statewide average weekly wage.

The Presidential task force report estimated that a worker totally disabled at age twenty-eight would suffer a lifetime economic loss of $151,152. In Pennsylvania, at the maximum weekly rate of $60 for total disability, he would recover only 39.5 percent of his lost income. In New Jersey, the same worker would recover 54.7 percent of his loss.

The task force was unequivocal in its findings:

> In short, we conclude that benefits paid in workmen's disability income programs are *inadequate*. We would argue that they should replace a high proportion of the income loss of the disabled worker. Evidence indicates that they are far from meeting this objective. They even fail to meet the more modest objective of the 66⅔ percent of wages set by the framers of the workmen's compensation program, and if the disabled worker and his family must look to any single program benefits as his exclusive income source, he will find himself living in poverty.[38]

The unwillingness of state legislatures to increase maximum weekly compensation benefits to keep pace with rising wages and living costs is a tribute to the political power of corporate interests and their allies, who have managed to eviscerate the state compensation systems while at the same time objecting to any sort of federal intrusion into their private preserves. The extreme lengths to which they were capable of going to defend the sorry status quo came to light in 1955, when Dr. Arthur Larson, an Undersecretary of Labor in the Eisenhower administration and a distinguished authority on workmen's compensation, undertook to prepare a model compensation act, which he stressed was intended only to help the states upgrade their own acts, and was not in any way the forerunner of any attempt to federalize workmen's compensation. His explanations were in vain. The model act stirred up a hornet's nest of opposition from business groups. Larson was bitterly attacked, and the Appropriations Committee report actually forbade the Department of Labor to continue work on the model act. The project had to be terminated, and mimeographed copies of the act became collector's items.[39]

An even more grotesque example of the distorted priorities set by business groups was their concerted effort to amend the Social Security Act in the early 1960's. Under the act, a person permanently and totally disabled from any cause could collect benefits. Thus, a worker whose permanent, total disability was job-related might be able to collect from two sources: social security, and workmen's compensation. The fact that in some cases a worker might receive from these two sources disability benefits greater than his pre-accident wage horrified the National Association of Manufacturers, the U. S. Chamber of Commerce, the American Bar Association, the entire insurance industry, and at least nine state legislatures—who, in a cloudburst of hypocrisy, lamented that this would be a disincentive to rehabilitation and therefore lobbied furiously to have the Social Security Act amended. These groups showed no such concern for the many inequities in the existing state compensation statutes, such as the maximum limit that some of the states put on medical benefits. Instead, they set out to pick the pockets of paraplegics and other seriously disabled workers. What made these efforts even more bizarre was the fact that less than 2 percent of those entitled to social security disability benefits were also receiving workmen's compensation, and less than .1 percent of *all* workmen's compensation awards involve permanent, total disability.[40] In 1965, their amendment passed—reducing social security benefits to the permanently, totally disabled employee so that the combination of social security and workmen's compensation benefits the worker was entitled to receive would not exceed 80 percent of his average earnings prior to his disability. The U. S. Supreme Court recently decided that this social security offset clause is constitutional.[41]

Having sketched the historical background of efforts to make American factories and workshops reasonably safe and healthy, we turn to a more detailed examination of the activities of various groups that have assumed responsibility for the protection of the workers. State governments, the federal government, and the labor unions have all exerted themselves on behalf of occupational health and safety. A progress report is now in order.

4

A Pattern of Neglect

Ever since 1877, when the Massachusetts legislature passed this country's first work safety law, state governments have involved themselves, for better or for worse, in the struggle to protect the safety and health of American workers. The states have set up the "front line" for the legal defense of workers against job accidents and diseases.

Industry, of course, would have preferred to handle job safety and health by itself, without outside interference, on a voluntary basis through the mechanism of consensus standards and voluntary compliance. As the reality of the human toll exacted by industrial growth made this proposition less and less tenable, business spokesmen grudgingly conceded that perhaps government did have a limited role to play, but only on the state and local levels.

Regulation of job safety and health by the states offered a number of theoretical advantages to the worker. Local officials understood local conditions and should have been in an advantageous position to gather comprehensive data on work injuries and diseases and their causes. The various states could have innovated with health and safety standards as well as methods of enforcement, in efforts to find optimum levels and approaches. Local government officials should have been responsive to the physical needs of the workers, who made up an important segment of their constituency. The states have traditionally regulated the compensation of work accidents and diseases, and should therefore have been vitally concerned with their prevention.

By the late 1960's, the moment was ripe, if not long overdue, for a thorough examination of the states' performance, to determine the extent to which the states had taken advantage of these opportunities.

Occupational safety

Key factors in evaluating the effectiveness of state efforts to reduce the frequency and severity of work accidents include: the percentage of workers covered by the state laws; the amount of money allocated by the states to the fight against work accidents; the legal tools that the states have given to their job safety personnel to safeguard the workplace; and the manner in which the states have enforced their job safety laws and regulations.

In 1969, the U. S. Department of Labor's Bureau of Labor Standards estimated that 9.8 percent of the total American work force (excluding private household employees) fell completely outside any sort of safety protection afforded by state authority.[1] This, of course, did not mean that 90.2 percent of all workers were adequately protected, but only that state laws created the possibility that they might receive some sort of protection in the form of state safety rules. But for 9.8 percent, there was not even the pretense of protection.

For example, in 1969, of the more than 700,000 men and women employed in Colorado, state law provided for the establishment of safety rules covering only 1.9 percent of them. The only workers protected in Colorado and Missouri, as well as in New Mexico and South Dakota, were those engaged in mining. Texas, with over 3.5 million workers, did not enact an occupational safety statute until 1967.

The most glaring gap in coverage left the agricultural workers of twenty-eight states without any legal protection against work hazards.[2] This omission, a monument to the political power of agribusiness, bears fruit each year in the job-related deaths of at least 2,500 farm workers, and the more than 210,000 disabling injuries suffered by these unfortunate laborers.[3] In addition, many thousands are exposed to toxic pesticides and other agricultural chemicals.

Gaps in coverage are but one indication that few states are seriously committed to the cause of job safety; a far better indication is the miserly allocation of men and money by the states to promote job health and safety. A twenty-five-state sampling by the AFL-CIO in 1968 revealed the priority lawmakers in these jurisdictions placed on human life: they employed one and a half times as many game wardens as safety inspectors.[4]

A Bureau of Labor Standards statistician has estimated that the ideal ratio of safety inspectors per worker is 1:25,000.[5] A survey by the bureau for fiscal year 1968 reported that Delaware had one inspector to cover a work force of 207,000.[6] The ratio in Michigan was 1:111,000, and in Indiana 1:90,600. In Mississippi, New Hampshire, and South Dakota there were no inspectors. Finally, it should be noted that most state safety personnel are assigned to inspect only elevators and boilers.

The bureau survey also listed budget allocations. Oregon, with the best budget level, allocated over $1 million to safety but employed only seventy-five inspectors. Wyoming, with the worst level, allocated $500 (exclusive of salaries) and employed three inspectors. This amounted to $2.70 per nonagricultural worker in Oregon, and less than $0.01 per worker in Wyoming. The average for the thirty-eight states surveyed was slightly more than $27 million for safety, or $0.48 per nonagricultural worker.

One would expect that inspecting factories for safety violations would be a full-time job, but Indiana's twelve inspectors must not only cover 6,609 workplaces in that state but also engage in all kinds of miscellaneous duties.[7] They are responsible for the prevention of occupational diseases. But they also represent the Governor on committees established to set prevailing wage rates for building tradesmen on public construction projects, and they investigate and make recommendations on applications for permits for lunch periods shorter than the amount of time required by law.

Indiana workers have paid for this misallocation of priorities with their lives. A recent coroner's report on the death of three steelworkers at a U. S. Steel plant in Gary amounts to a startling indictment of the shortcomings of the Indiana factory inspection system:

> It is conceivable that if the state had employed persons with sufficient training coupled with a method to guarantee periodic inspections to large industrial plants, this accident might have been prevented. In this case, not only was the state's safety inspector ill-equipped to do his job, the Division of Labor had not inspected this plant during the entire time of its operation prior to the deaths on January 11, 1970. It is obvious that the number of inspectors and the degree of training which they have make the system of safety inspection in Indiana totally inadequate.[8]

Restrictions on enforcement powers further weaken the protection afforded by state safety laws. A 1969 Department of Labor survey found that in twenty-one states inspectors lacked legal authorization to shut down machinery on the spot when they found a particular work environment imminently hazardous.[9] Over 20 million workers in these states could expect no immediate help from the inspectors in the face of serious danger to life and limb.

Sixteen states provided no criminal sanctions against employers who deliberately violate safety laws,[10] even though such violations might expose employees to the risk of grave harm or death. Thus the law winked at this brand of willful misconduct on the part of employers. In five states safety inspections were an obvious charade since inspectors did not even have the right to enter work premises.

In Ohio, the Industrial Commission has engaged in a bit of legalistic flimflam to place further restrictions on its authority to promote work safety. The Ohio constitution states that an employee can obtain increased workmen's compensation benefits if he is disabled as a result of his employer's violation of a safety regulation specifically worded to cover enumerated work hazards in a particular industry. Although this provision refers only to increased compensation benefits and says nothing about accident prevention, the Ohio Industrial Commission inexplicably reads it as restricting the commission's authority to draft safety codes.[11] According to this interpretation, the commission must draft specific codes for specific industries and cannot draft more generalized safety regulations.

The commission has drafted and promulgated nine codes setting certain specific safety standards for nine industries. But because of its self-imposed restriction, it has not fashioned standards that might be applied across the board to all workers. Exposure to toxic substances and noise can harm any employee regardless of the specific job he is doing, and it would be relatively quick and easy for the commission to set general limits to such exposures. But the commission has chosen not to protect Ohio workers from these hazards.

The various inadequacies of state safety laws and regulations tell only part of the story. A close look at how these laws and regulations are enforced suggests a shameful pattern of non-enforcement.

The states have been understandably reluctant to publicize their enforcement records. Although it would be easy to compile records of the number of prosecutions and penalties assessed for safety violations, few states bother to do so. New York reported only 6 fines in 1968, although employers committed over 10,000 violations, and the State Division of Inspection and Safety referred 442 cases for prosecution.[12] This no-prosecution policy became a political issue in the 1970 campaign for the office of Attorney General in New York. Democratic candidate Adam Walinsky filed suit to remove the state's Industrial Commissioner for nonenforcement of the law, charging that violations of state safety laws contributed to the deaths of fifty-eight workers in 1968.[13] Walinsky claimed that his opponent, incumbent Attorney General Louis Lefkowitz, should have ordered prosecutions in at least fifty-two of these cases. Out of 24,845 violations in Massachusetts in 1968, there were 28 prosecutions. Twelve fines were assessed, for an average of eighty-eight dollars.

According to a survey by the International Association of Industrial Accident Boards and Commissions (IAIABC), in only 25 percent of those states that give safety inspectors "redtag" authority was this power invoked one time or more to close down a work environment deemed imminently dangerous.[15] The remaining 75 percent reported no use of this provision.

In 1965 the U. S. Department of Labor's Bureau of Labor Standards considered a proposal to make an enforcement survey.

Necessary information would have been supplied by the states on a voluntary basis and would have constituted invaluable data highly relevant to the Congressional hearings on occupational safety and health in 1968, 1969, and 1970. The Bureau did not go ahead with the survey because, in the words of a bureau employee, Thomas Seymour, "It wasn't politically good at that time. Also, the Bureau of the Budget has the authority to approve forms if more than ten people are to be interviewed, and our survey form was not approved—too much red tape."[16]

The 1970 IAIABC survey,[17] which obtained responses from all but ten states, made only a feeble attempt to obtain information about enforcement procedures. The survey confined itself to a compilation of the number of safety-violation hearings held but made no effort to gather data on individual violations or prosecutions. Since these hearings range from formal procedures to informal "let's-talk-it-over" sessions, mere compilation of the number of these hearings tells us very little about enforcement.

The real weakness in the administration of state safety codes stems from the philosophical approach to enforcement taken by virtually all the states. The strange notion has become deeply rooted that corporate lawbreakers should not be penalized, but merely warned, in order to give them the opportunity for "voluntary compliance" with safety regulations. Perhaps this attitude in part reflects the lack of money and manpower that the states allocate to work safety, and the concomitant hope that a policy of encouraging voluntary compliance will somehow offset this inadequacy. The record indicates that this hope remains unfulfilled. Employers have little incentive to take any initiative to root out unsafe work conditions and practices. Instead, they can subject their employees to all kinds of hazards, to be corrected only if discovered by state inspectors—if and when the plant is visited.

A revealing illustration of the type of enforcement this attitude breeds is provided by the following "Case Study of a Chronic Violator," compiled by Dr. George Hagglund of the School for Workers at the University of Wisconsin:

> This particular company "has been a chronic violator in just about every health and safety code in the book since the early 1950's," according to a source in the [Wisconsin] Divi-

sion of Industry, Labor and Human Relations. It has been cited for violating the health and safety codes, building codes, fire regulations, pollution regulations, and the child labor laws. The individual violations number in the hundreds, but the company has paid little in fines and costs over the years. Three recent safety cases involving this company illustrate the way in which the company has been allowed to continue as a chronic violator.

The First Case (1961)

An inspection of this plant was made on August 14, 1961, and nineteen separate violations of the Industrial Commission's Safety Orders were discovered. Most of these violations involved the company's failure to provide guards for machinery and guard rails for open pits. The company was given until December 1 to correct these violations, but refused to allow Inspector Sigurdson entry to reinspect the plant on November 27. When a reinspection was finally made in late January, 1962, *more than five months after the initial inspection*, the company had corrected only four of the original nineteen violations. Another reinspection was made on April 4, and only eight violations had been corrected by then. By the end of 1962, all the safety violations had been rectified. The firm was fined two hundred dollars plus court costs, and the case was closed.

The Second Case (1966)

The Industrial Commission left the company alone until April, 1966. On March 23, the local volunteer fire department sent the commission a strong letter urging them to inspect the plant, stating:

> "We, the thirty members of the Merton Volunteer Fire Department classify this factory and all its operations as a public nuisance. They violate all the health and safety rules in the book. It is high time a complete inspection of this factory is made by your department and we request that this be done immediately."

Copies of the letter were sent to both U.S. Senators, the state Fire Marshal, and the appropriate U.S. Representative, State Senator, and State Assemblyman. This approach got results, for the Industrial Commission inspected the plant on April 15 and found seventy-seven violations. A compliance date of July 1 was set at that time. The company made little effort to rectify the situation, and the case was given to the Attorney General for prosecution on November 22, 1966. The proprietor wrote Governor [Warren] Knowles, asking him "to get the Industrial Commission off my back," but Knowles refused, criticizing the firm for its uncooperative attitude and activities. A reinspection on November 14, 1967, *nineteen months after the initial inspection*, revealed that only forty-six of the seventy-seven violations had been corrected. Additional inspections were made on January 2 and February 12 of the following year. All the violations were corrected by March 13, 1968, almost two years after the initial inspection.

This case went to the Attorney General's Office in November, 1966. A complaint was prepared asking for $14,600 in forfeitures. The case was transferred in September, 1967, to the County Corporation Counsel for the county involved. Two pretrial hearings were held in late 1967 and the reinspections revealed that the company was slowly complying by the early part of 1968. In February, 1968, without consulting either the Attorney General or the County Corporation Counsel, the regional supervisor of the Industrial Safety Division informed the defendant that the action would be dropped. *There were no penalties or court costs assessed against the company.*

The Third Case (1969)

A third safety case was recently brought against the same firm and is still pending. On May 6 and 20, 1969, another inspection was made of the same plant and a total of twenty-six violations noted. Twenty of these were safety violations, and the others were building code violations. A reinspection was made on September 1, and only ten of the twenty-six violations had been corrected at that time. The compliance date was extended to December 1, but another reinspection was not made until January 6, 1970, *eight months after the initial inspection*. At that time, the company had corrected only half the violations. The case was given to the Attorney General for prosecution on February 20, 1970, and it is still pending. The state of Wisconsin is seeking $2,800 in forfeitures.

Conclusion

Wisconsin law provides that an employer violating safety and building codes can be fined amounts ranging from $10 to $100 per day for each code violation. In the preceding case, the Attorney General's Office sought fines and costs amounting to $17,400 in the 1966 and 1969 cases and an unascertained amount in the 1961 complaint. Thus far the company has actually paid $200 in fines over a ten year period! If chronic violators can get by with financial damages of $200, how hard-hit are companies with better records, but still in violation of safety and building codes? The handling of this case emphasizes the importance of interested parties filing safety complaints with the Safety and Building Division and following through to see that investigations and compliance is secured.

In a letter to a member of the Study Group, Dr. Hagglund described how difficult it was to put the case study together:

We ran into considerable difficulty putting even this amount of information together. We were told by an Assistant Attorney General that no single agency possesses a full file when a case goes from the Building and Safety Division, to the Attorney General's Office, to the Waukesha County Corporation Counsel for trial. It seems that files are thrown away by the court after the case has been processed, and other agencies only have the correspondence related to their end of the case. This makes difficult the systematic research job necessary to document performance of a Wisconsin state agency.[18]

Enthusiasts of the voluntary compliance approach place much stock in convincing companies that "safety pays." The Wisconsin experience undercuts this slogan, teaching that it does not hurt a company financially to maintain unsafe working conditions.

A recent disclosure in Washington, D.C., revealed another ploy companies can use to make a mockery of the "safety-pays" motto. Charles Green, the District of Columbia's Director of Industrial Safety, announced that in 1970 companies accused of willfully violating work safety regulations saved seventeen thousand dollars by forfeiting collateral rather than appearing in court to answer charges.[19] Collateral forfeitures cost these companies twenty-three thousand dollars. If they had gone to court and had been forced to pay a minimum fine of one hundred dollars per day for each infraction, they would have paid out a total of forty thousand dollars. The imposition of fines up to the maximum limit of six hundred dollars per day would have proportionately increased this total.

Employers who find it distasteful to be caught violating the safety codes, even at bargain rates, often have the chance to clean up their plants in advance of inspections, since some states are said to tolerate, if not actually authorize, a policy of notifying companies in advance that the safety inspectors are coming. Union officials charge that this practice in New Jersey makes a farce of inspection. Frank Sass, union safety chairman in a New Jersey factory, has complained, "We also have the problem of the state inspector coming in on request of the company . . . the man is there at their request, and they will show him around. Naturally they show him only what they want him to see."[20] In Ohio, even union requests for inspections are funneled through management. William Gaines, service representative for the United Auto Workers in a Cincinnati factory, has stated: "We have a system where if you have a complaint you call the state offices. Then they advise the plant, and by the time the inspectors get there there are no more safety problems, they have done a good job covering up the problems."[21]

Although the states have displayed little enthusiasm for enforcing job safety laws strictly, they have not hesitated to stress worker education as a means to prevent work accidents. This represents the triumph of attitudes such as those of a National Safety Council official who declared in 1933:

> If I may be permitted to criticize any of the states in the United States, I think that criticism should be leveled against those states which have devoted too much time and attention to the enforcement part of their programs. Safety can never be legislated and enforced into individuals. Safety must be sold and taught into individuals.[22]

The approach followed in Ohio typifies this philosophy of accident prevention through education. *The Monitor*, the official publication of the Ohio Industrial Commission, devotes substantial space to the announcement of safety awards. The commission designated 1969 as "19-Safety-9" and sought to reduce the industrial accident toll by popularizing silly slogans such as:

> As the year commences
> We can establish defenses
> Against safety offenses
> By sharpening our senses.[23]

The commission does not have to rely upon safety jingles to reduce work accidents and diseases in Ohio. A 1923 amendment to the Ohio constitution established a special fund, derived from a percentage of each employer's workmen's compensation insurance premiums, and specifically earmarked it for safety and hygiene inspection. It strains credulity to discover that the commission has not only failed to spend all the money in the fund, but actually passes the savings back to the employers! Currently there is a surplus of $7.2 million in the fund.[24] From 1963 to 1967 less than $1.7 million per year was taken from the fund and used for safety and hygiene inspections. At the same time the contributions of employers to the fund have been reduced from 1 percent to .75 percent of the premiums paid for workmen's compensation insurance.[25]

Mr. Holland Krise, chairman of the Ohio Industrial Commission, explained in a letter to the Study Group that, "Due to increased payrolls resulting in larger premium income, it was found that the trust fund was increasing without justifiable means of spending it. Therefore the commission reduced the amount to be collected from 1 percent to .75 percent." Thus, the commission could not find any "justifiable" reasons for spending this money to help safeguard Ohio workers against the risk of industrial accidents or diseases. The Ohio AFL-CIO has suggested that, at least in 1963, the commission had little difficulty in finding uses for the fund. According to the union magazine, *Focus*:

> Industrial Commission members raided the safety fund to air condition their offices. It was also this year that the same fund was tapped for $90,000 to help pay for Governor James A. Rhodes' "Wonderful World of Ohio" booklet. And $6,268.12 was paid from the fund for a special survey on industrial safety conducted by the son of one of the commissioners at that time. Money was also spent to pay two state legislators for delivering lectures on industrial safety.[26]

Total preoccupation with the voluntary compliance approach to industrial safety emanates from the very pinnacle of the safety establishment. In the words of Howard Pyle, president of the National Safety Council:

> With all due respect to the desirable effects that the promulgation and enforcement of standards can have on the nation's

occupational safety and health performance, we must recognize that the successes that have been achieved so far are largely the result of the dissemination of safety information, the implementation of proven countermeasures, and education and training of employers and employees.[27]

In the crucible of reality, this approach represents the triumph of massive insensitivity. It has not blinded every state administrator. At the 1969 IAIABC convention, a Florida official observed:

For twenty-one years following the establishment of the Department of Safety in 1934, education was our only tool. Regrettably it took the most tragic and costly accident in the history of Florida industry to spur our legislature toward safety regulations.[28]

(The tragedy occurred when an elevator hoist at a Jacksonville construction site plunged five stories. Seven workers were killed, and twelve others were permanently disabled. One of the victims died in 1967, ten years after the accident, during which time he had existed in a state of unconsciousness.)

Occupational health

If state occupational safety programs are inadequate, they still look good when compared with the feeble efforts the states have made to combat industrial diseases. To begin with, although every state has some form of job safety regulation, as of 1968 eight states had no recognizable industrial health programs at all.[29] Indeed, this represents backsliding, for in 1950 only three states lacked job health programs. State budget reductions and a cut-back in federal grants-in-aid for industrial hygiene under the Social Security Act caused this regression. The record compiled by the states does not derive from the novelty of the notion of occupational health, since both New York and Ohio initiated such endeavors more than half a century ago.

Health programs administered by the states have turned out to be totally incapable of protecting workers from industrial diseases. As manufacturing processes have become more complex, and the range of products more diverse, the worker is left to face the onslaught of subtle, silent, and at times deadly chemicals, gases, dusts, and fumes. As modern technology creates and compounds industrial health hazards, state occupational health programs have been worse than useless—for they present a facade which can lull unsuspecting workers and their families into believing that there is some sort of effective government regulation over the health hazards of the workplace, when in fact employees are completely at the mercy of their job environment and its corporate creators.

The scope of the occupational disease problem is dramatic evidence of the overall failure of the states. It would be useful to discuss briefly some of the particular shortcomings as they relate to staffing, standard-setting, data-gathering, and organization.

In 1969, C. C. Johnson, Jr., administrator of the HEW Consumer Protection and Environmental Health Services, stated, "There is an immediate need for approximately twenty-eight thousand additional industrial hygienists, physicians, nurses, and scientists to protect adequately the nation's workforce."[30] Statistics on the ratio of occupational health professionals to the working population point up the scarcity of health personnel. HEW's Bureau of Occupational Safety and Health has estimated that an optimum ratio would be one professional per 35,000 workers.[31] A 1968 study revealed that in twenty-four states, which employ 36 percent of the work force, there were from two to nine health professionals, making a ratio of one professional for every 152,-000 workers.[32] Eleven states had ten or more professionals; in these, the ratio was 1:92,000. Nationwide the ratio was one professional for every 121,000 workers.

Over the past ten years the number of physicians employed by state occupational health units has dropped from fifty to thirty-five. Five of the fifteen who left had been program directors. The surviving thirty-five comprise part-time and full-time, as well as consulting, physicians. In the past ten years the number of nursing consultants has dropped from thirty-six to twenty-two, five of whom are part-time. Only twelve states and six local units employ these twenty-two nurses. This is a significant statistic, since the decline in the number of nurse consultants leaves the majority of states without any way to advise and communicate with in-plant nurses, who are employed by industry and provide most of the on-the-job occupational health services currently offered to workers.

The expansion of environmental services by the states has diverted personnel away from the specific field of occupational health. In eighty-two state and local units dealing with occupational health, ten agencies are one-man outfits, and only twenty units in eleven states employ more than ten people.[34] But in these latter twenty units, 40 percent of the positions are designated for nonoccupational radiological health, air pollution control, government-employee health services, and other miscellaneous functions not directly related to occupational health. Of the personnel increases in state health agencies over the past twenty years, half were allocated to environmental services not directly concerned with occupational health.

One of the primary tasks of these small staffs is to set standards of exposure to occupational health hazards. The performance of the states in setting maximum allowable exposure limits has been shameful. Most states tend to rely upon the American Conference of Governmental Industrial Hygienists (ACGIH), a private standard-setting organization. ACGIH has adopted so-called Threshold Limit Values (TLV's), setting maximum exposure limits over an eight-hour period for only 450 chemicals and compounds,[35] leaving exposure to thousands of substances unregulated. And many TLV's, which the states tend to accept without independent analysis, are far too high, as shown in Chapter 2. Those states that adopt and use nationally recognized standards often allow them to become outdated. ACGIH reviews and revises certain of its standards each year on the basis of new information. Only three states kept the standards they had borrowed from ACGIH up to date as of 1969. The remaining states have standards that are from three to twenty years old.

Other states adopt their own standards and set them in several ways. They may promulgate statutory or administrative standards that set quantitative limits on exposure. Or they may permit "reasonable" exposures to gases, dust, etc. Quantitative standards put employers on notice about their exact legal responsibilities and also warn workers of amounts of exposure that may be dangerous to them. Hence they are far more effective than standards that speak of "reasonable" exposures, or exposures that are "not injurious." Nonetheless, only thirty-three of the seventy-two state standards on dust in force in the late 1960's were quantitative, while the remaining thirty-nine used vague language.[36] Only fourteen states regulate noise, and of these, only six use quantitative standards. In order to develop their own standards, state occupational health departments need basic information about the nature and amount of harmful exposures experienced by workers. Yet state officials have not displayed the initiative and enthusiasm necessary for investigative work that might reveal the causes of industrial diseases.

At this time, no state knows the full extent of its occupational health problems. One glimpse at the enormity of this information gap came to light in 1966 as a result of a survey conducted by the state of Florida in cooperation with HEW's Bureau of Occupational Safety and Health. A report on the survey found that:

> Knowledge of health hazards was minimal, as was evidenced by the fact that management in 88 percent of all establishments did not feel they had any health hazard, although 11 percent of the establishments were rated as high priority, i.e., needing industrial hygiene attention within one year. Ninety-nine percent of all establishments did not employ a physician on a regular basis, and 18 percent had no nursing service available. Safety programs did not exist in 68 percent of all establishments.[37]

The results of a Bureau of Occupational Safety and Health survey in Chicago revealed strikingly similar results. The study found the level of medical assistance as bleak as that found in Florida, and also revealed that 50 percent of the Chicago plants did not receive help in any phase of occupational health from insurance companies, corporation headquarters, or state agencies.[38] The study made no estimate of the number of industrial hygienists Illinois would need. In 1968 when the survey was taken, however, the state of Illinois employed two hygienists for a working population of 4.7 million. The situation in Chicago is further explained by the fact that, according to another Bureau of Occupational Safety and Health Department study, the city's acting commissioner of the Health Department in 1969 had "a distinction and disqualification for his present job deriving from the fact that he was Mayor Daley's family obstetrician."[39]

If state occupational health directors have no idea of the prevalence of industrial diseases in their jurisdictions, they cannot be expected to make a convincing argument to their legislatures that their health programs merit additional funding, or additional authority to promulgate and enforce health standards.

Physicians could furnish the state agencies with valuable data on occupational diseases, but almost all the states have failed to take best advantage of this important source of information. The so-called first-report system has been effective only in California, where the law requires the physician to send to the Department of Industrial Relations a form virtually identical to the form he submits to the workmen's compensation insurance carrier for payment for his services. Other states require employers to report occupational disease disabilities for which *workers* receive compensation. But since many diseases are not covered by state workmen's compensation statutes, this approach is hardly effective.

Making matters still worse is the fractionalization of authority among various departments within state governments, debilitating occupational health programs and making it all but impossible for the states to mount a sustained, coordinated attack on the increasing health hazards of the workplace.

In Connecticut, the Health Department has statutory authority to administer an occupational health program and to record industrial diseases as they are contracted, while the Department of Labor has the right to enter plants to make inspections.[40] In California, two different divisions of the Department of Labor have the right of entry and responsibility for occupational disease reporting.[41] In only fourteen states does the health department have a specific right of entry (and, characteristically, only three of these jurisdictions provide a legal sanction against employers who refuse to admit inspectors from the health department).[42]

The job safety and health record compiled by the states, and described above, fully justifies the perspicacity of industry officials who at an early date decided to use state regulation as a fall-back position in their strategic retreat from public involvement in the fight against occupational accidents and diseases. The states refused to capitalize on the opportunity they have had over the past century to develop vigorous and effective work safety and health programs. Instead, they have allowed state regulation to be used as a barricade behind which industry could wave the banner of "states' rights" and resist federal encroachment into the field of occupational health and safety.

Industry's judgment that in the long run health and safety regulation on the state level would prove beneficial to corporate interests rested on the assumption that it would be easy to convince state legislatures to enact job safety and health laws that would be acceptable to business, and to convince administrators to apply these laws leniently. The threat was implicit that tough laws and vigorous enforcement might produce corporate decisions to relocate existing plants to other states, and to build new plants elsewhere. In addition, as industry grew and spawned exceedingly complex health and safety problems, the individual states would not have the financial resources to deal with them.

The corporate strategy to keep government regulation of work safety and health in the hands of the states succeeded until the late 1960's, when Congress finally held extensive hearings on the occupational environment. Then the price American workers were paying for this lack of effective legal protection became painfully evident, and the issue before Congress quickly became not whether to enact a federal statute, but what kind of federal statute to enact.

Indeed, regard for state regulation was so low that former Secretary of Labor W. Willard Wirtz, in response to a suggestion that his department should have consulted state officials in the course of drafting a federal job safety and health law, replied, "We should ask the foxes what we should do to the chicken fences?"[43]

5

Pennies and Posters

The belief that protecting the work environment was primarily the responsibility of the states as a matter of "states' rights" prevailed until the late 1960's, a triumph of ideology over humanity. For decades Washington meekly accepted a minor, supporting role in the public regulation of job safety and health, providing research and some technical assistance through the U. S. Public Health Service.

There have been several exceptions. One is federal employment itself. Another is railroading, an interstate commerce that did not lend itself to piecemeal regulation by the individual states. Railroad safety became a matter of Congressional concern as early as 1893, when the first federal law requiring safety equipment on engines and cars was passed. The federal government also has had jurisdiction over maritime matters, but only in 1960 did the Department of Labor finally promulgate comprehensive safety and health regulations to apply to the shipyard and longshore industries. In 1910 Congress created the Bureau of Mines in the Department of the Interior, in response to a series of mine disasters, but gave the new agency no authority to inspect mines or to enact or enforce safety and health standards. It was not until 1941 that the bureau received authority to make inspections and investigations in the mines. In 1952 Congress at last gave the bureau authority to promulgate and enforce standards, but only to prevent "major disasters," a serious limitation on the bureau's work. Finally, the federal government has engaged in several programs aimed at protecting the safety and health of employees in the private sector working on certain types of government contracts.

The Study Group has chosen for study three examples of federal involvement: the Bureau of Occupational Safety and Health (hereinafter referred to as BOSH and today known as the National Institute of Occupational Safety and Health), a subdivision of the Department of Health, Education and Welfare—charged with supplying technical assistance to public and private groups engaged in efforts to improve levels of job health and safety; the Department of Labor's performance under the Walsh-Healey Public Contracts Act, which directs the Department to set and enforce safety and health standards for employees working for private employers on certain contracts between the employer and the federal government for the manufacture or furnishing of supplies; and the efforts of the federal government to reduce the accident rate among federal employees.

Bureau of Occupational Safety and Health (BOSH)

To say that the Bureau of Occupational Safety and Health has been lost somewhere in the basement of the federal bureaucracy is an understatement. The following exchange between Dr. Marcus Key, director of BOSH, and Senator Harrison Williams during the 1970 Senate Hearings on Occupational Safety and Health, records the Senator's valiant attempt to locate BOSH.

> SENATOR WILLIAMS: You are Dr. Key?
> DR. KEY: Yes.
> SENATOR WILLIAMS: And your job is?
> DR. KEY: Director of the Bureau of Occupational Safety and Health.
> SENATOR WILLIAMS: . . . you are the boss-man in charge of this; am I right?
> DR. KEY: That is correct. . . . We are one of five bureaus in the Environmental Control Administration.
> SENATOR WILLIAMS: In the what?
> DR. KEY: The Environmental Control Administration, which is one-third of the Consumer Protection and Environmental Health Service. . . .

SENATOR WILLIAMS: All right. Here we have on top, in this area, what is it, "consumer?"

DR. KEY: Consumer Protection and Environmental Health Service, which in turn is one-third of the Public Health Service.

SENATOR WILLIAMS: All right. And then you are one division under Consumer Protection: Occupational Health and Safety?

DR. KEY: Two levels. The Environmental Control Administration with the other bureaus, Community Environment Management, Radiological Health, Solid Waste Management, and Water Hygiene, is our immediate home.

SENATOR WILLIAMS: Wait a minute. You left me at the gate. I thought you were Occupational Health and Safety?

DR. KEY: Yes, sir.

SENATOR WILLIAMS: Where do you get into solid wastes?

DR. KEY: Well, the environment programs have been grouped together, in the Environmental Control Administration. We take care of the environment inside the factory. These other bureaus I mentioned take care of the environment outside the factory.

SENATOR WILLIAMS: I see.[1]

In fact, in 1968 BOSH was part of the Environmental Control Administration, which was part of the Consumer Protection and Environmental Health Service, which in turn was part of the Public Health Service, a division of the Department of Health, Education and Welfare. When one takes into account the Office of the Undersecretary for Health and Scientific Affairs, which had been interposed between the Secretary of HEW and the Public Health Service, it turns out that four levels of bureaucracy insulated BOSH from the Secretary of HEW.

In testimony at the 1969 House Hearings on Occupational Safety and Health, Bill B. Benton, Jr., formerly BOSH's program officer, recounted the downgrading of BOSH from the pre-1967 period, when only two levels of bureaucracy separated the bureau from the Secretary. He noted that "occupational health is not only at the bottom [of the bureaucratic ladder] but is going down."[2] Before 1967, what is now BOSH was known as the Division of Occupational Health, part of the Bureau of State Services, which was part of the Public Health Service. In 1967 it became part of the National Center for Urban and Industrial Health, and moved to Cincinnati. In 1968 it became BOSH and returned to Washington as part of the Consumer Protection and Environmental Health Service. Shortly thereafter, another yo-yo-like move sent BOSH back to Cincinnati.

This bureaucratic shuffling supports a conclusion that is easy to draw from a perusal of Congressional appropriations hearings during the 1960's: the federal government felt no sense of urgency or priority about the problems of job safety and health. Testimony by government officials on behalf of the occupational health program was invariably tepid and, with but one minor exception, unsupported by nongovernmental groups such as labor unions or professional groups.

It should not be assumed, however, that these government officials were unaware of the dimensions of the occupational health problem and of the potential role that could be played by their agency. In 1965 BOSH's predecessor, the Division of Occupational Health, submitted to the Surgeon General the so-called Frye Report entitled, "Protecting the Health of Eighty Million Americans: A National Goal for Occupational Health."[3] This document clearly recognized the dangers posed by new industrial chemicals, the continuing human toll taken by known occupational diseases, and the growing suspicion that many chronic illnesses such as cancer and heart disease might be causally related to work exposures. The report outlined a new national program and recommended a budget of $50 million, a far cry from the division's budget of $6,592,000 for fiscal year 1967 (although still hardly adequate to cope with the nation's job health problems). But occupational health officials were unable to mobilize any effective political support for this program.

Prior to 1968, responsibilities for occupational safety and health were kept separate, with the latter firmly in the grasp of physicians from the Public Health Service. When the Division of Occupational Health became BOSH in 1968, the physicians remained in control, and, in the light of severe funding limitations, it is not surprising that they paid little attention to safety.

The functions assigned to BOSH have always been broad and of vital significance in the campaign for a safe and healthy workplace. In 1968, for example, the Secretary of HEW gave BOSH the following mission:

Plans and conducts a comprehensive program of demonstration and technical assistance to governmental agencies, professional groups, universities, labor, industry, and health-related organizations for the improvement of the health and safety of the working population. Conducts studies, field investigations, and demonstrations for the detection and control of occupational and occupation-related disease and injury. Collects and analyzes data on the health status of the working population and on industrial materials and processes currently in use. Develops criteria for standards for safe and healthful working environment. Directs studies on the economic and social consequences of occupational and occupation-related diseases and injuries and promotes occupational safety and health programs in state and local agencies.[4]

Yet BOSH has always been so pathetically underfunded— its 1968 appropriation was only $6,667,000—that it is difficult to see how the Secretary of HEW could list all these responsibilities with a straight face. Indeed, Mr. Benton disclosed that in fiscal year 1968, BOSH was not even allowed to retain the $4,544,000 appropriated to it for direct occupational health operations, for $1,219,000 was "tapped" from it and diverted to help fund the National Center for Urban and Industrial Health, of which BOSH was but one division.[5] In 1969 the "tap" was increased by $300,000, wiping away an increase in the Congressional appropriation to BOSH.

Inadequate funding and a low sense of priority within HEW served to reinforce certain questionable attitudes assumed by BOSH personnel. One of the most unfortunate was a servile posture toward industry and management.

Since 1914, when the Office of Industrial Hygiene and Sanitation was organized in the Division of Scientific Research within the Public Health Service, the federal attack on occupational health and safety hazards has suffered from the absence of a broad legal authorization to enter workplaces to collect data necessary for meaningful studies. As a result, federal occupational health officials have had literally to beg industry for access to workers and records. During the course of time, this has spawned in federal officials the feeling that the best way for them to accomplish their mission is to avoid stirring up trouble and to defer completely to management's wishes.

To keep from rocking the boat, BOSH adopted a policy of not divulging the names of any plants or companies investigated by the bureau and found to contain occupational hazards. Even when BOSH officials knew that companies were not following their recommendations, with the result that workers were suffering constant exposure to known hazards, BOSH would not release the names of the companies to the public. Indeed, bureau people were reluctant ever to point the finger of blame at industry for occupational hazards.

An air of academic aloofness permeated the bureau. BOSH's general research findings were often delivered at professional meetings attended only by management and university people. Dr. Marcus Key, BOSH's director, recognized this as a problem in a paper delivered at a meeting of the American Public Health Association in Houston, Texas, in October, 1971: "Federal researchers have become an island of excellence, proud of their technical abilities and competence to research the problems and recommend solutions, but reluctant to join the firing line, where application of such knowledge is sadly lacking."[6]

The most damaging result of this attitude has been that BOSH personnel completely neglect to communicate with workers—the natural and necessary constituents of any occupational health program. Dr. Key himself reflected this attitude when he told a member of the Study Group, "I don't want to sit down with a bunch of guys and hear them rant and rave about the injustices suffered by workers."[7]

Dr. Hawey A. Wells, Jr., in a speech delivered at an Occupational Safety and Health Conference sponsored by the AFL-CIO on June 4, 1969, in Washington, D.C., forcefully criticized the attitude of BOSH personnel toward workers:

> I have been frustrated in my eleven-odd years in different capacities in the Public Health Service by that condescending attitude that many professionals and semiprofessionals there approached organized labor with. I believe that goes to the root of the problem.[8]

BOSH's refusal to develop a close relationship with the labor movement may have contributed to the bureau's lack of success in selling its program to Congress. Granted that the unions have, until recently, failed to push hard for job safety, health legislation, and appropriations; nevertheless, BOSH officials could have at least tried to woo labor. They certainly had nothing to lose.

One of the greatest obstacles BOSH has had to confront has been the lack of interest displayed by top officials of HEW in their own occupational health programs. When public opinion has been aroused over a particular occupational problem, such as air pollution or radiological health, the Department has removed the problem from the mandate of BOSH and established a separate unit to handle it. This has served to maintain the invisibility of occupational safety and health and to defuse in advance pressures that might be brought to bear on HEW to develop a strong, effective job-environment program.

Further evidence of the low estate of job safety and health within HEW came from the persistent refusal of the Department to seek statutory authority to gain a right of entry into plants working on contracts with the federal government, and hence covered by the Walsh-Healey Act. HEW made no attempt to negotiate with the Department of Labor for right-of-entry privileges under Walsh-Healey for BOSH or any of the bureau's predecessors.

Every BOSH employee with whom Study Group members talked said that one of the major problems that has hampered the bureau has been its lack of power to enter factories and workshops to collect data on workers and the workplace environment. This has been consistently cited as the reason occupational health people have had to be so solicitous of business and management. As Chris Hansen, commissioner of the Environmental Control Administration in HEW, said to the Study Group, "Our people have to go into a plant, bow from the waist, and beg." Yet HEW has never fought to correct this major deficiency.

Dedicated and hard-working though some BOSH officials may be, the fact remains that the bureau never really displayed the initiative and imagination required to promote the cause of occupational health within HEW, or before Congress. Its air of aloof professionalism helped keep BOSH buried in bureaucratic obscurity and, at the same time, failed to foster within BOSH personnel a sharply critical and independent attitude toward industry. From the point of view of job health, BOSH managed to wallow in the worst of both worlds.

The Walsh-Healey Act

In 1936 Congress passed what became popularly known as the Walsh-Healey Act, a piece of New Deal legislation that sought to regulate wages, hours, and conditions of labor for employees working on contracts their employers had signed with the federal government for the manufacture or furnishing of supplies and equipment in any amount exceeding ten thousand dollars. An employer found guilty of violating the act could be "black-listed" from federal contracts for a period of three years. The Secretary of Labor was given authority to promulgate rules and regulations necessary to carry out the provisions of the act. The Secretary was also charged with enforcing the act. Congress extended federal regulation of wages, hours, and labor conditions to cover service contracts with the federal government in any amount exceeding twenty-five hundred dollars under the McNamara-O'Hara Act of 1966, and federally funded and assisted construction contracts under the Construction Safety Act of 1969.

In 1968 Secretary of Labor W. Willard Wirtz called the Walsh-Healey Act "a program which has shown some progress over the years, but a program which we can't be proud of. . . . It is not the most effective kind of instrument."[9] At that time some 27 million workers were covered by Walsh-Healey, with another 6 million falling under the protection of the McNamara-O'Hara Act.

One of the main problems with these acts was that they affected only those employees doing work on federal contracts. Since part of the work force within a particular factory might be working on nonfederal business, they were unaffected by the federal statutes. Therefore, different safety and health standards might be applicable to different employees working within the same plant. This presents obvious enforcement difficulties.

Another big problem has been a consistent lack of adequate resources allocated to the Department of Labor for inspecting workplaces covered by Walsh-Healey and McNamara-O'Hara. In 1969 the Bureau of Labor Standards (hereinafter referred to as LSB) estimated that seventy-five thousand work establishments were covered by Walsh-Healey. Fewer than 5 percent of them were inspected that year.

The federal government actually made a more comprehensive effort two decades ago to enforce Walsh-Healey than in the late 1960's. There were about as many annual government inspections for $5 billion in contracts covered by Walsh-Healey in 1941 and 1942 as there were for $55 billion in the period between 1966 and 1968. This, of course, reflected a commitment on the part of the government to maintain and increase production levels during wartime. Industrial accidents would have hampered the war effort. It also demonstrates what the government can do when it sets a priority and allocates not only rhetoric but also resources.

The Congressional attitude toward Walsh-Healey in peacetime came to light during a Senate debate in 1952, as a result of an attempt to increase the Department of Labor appropriation in order to enable the Department to enforce properly all sections of the act, including the job safety and health provisions. Some of the Senators expressed the curious concern that the Department of Labor would send its inspectors into filling stations. Exclaimed Senator Lyndon B. Johnson of Texas:

> They are already in filling stations. They are going into small filling stations and saying to the individual station owner, "You are subject to this act. We are going to take jurisdiction, and we are going to prosecute you if you do not live up to our rules and regulations. . . ." If we raised the appropriation and gave them a few more "house mothers" there is no telling where they might go from the filling stations.[10]

By a voice vote the Senate rejected the proposed increase in the appropriation.

In its examination of Walsh-Healey, the Study Group attempted an in-depth study of the enforcement of the act, which is the responsibility of the LSB. But a major obstacle—the veil of secrecy surrounding inspection reports—hindered the effort. Instead of using inspection results as a compliance tool, the LSB had given orders that no information was to be released about the findings of any such inspection, even when the inspection had been initiated as a result of complaints by workers.

Indeed, the Study Group discovered that secrecy prevailed at every stage of the process. When a Walsh-Healey inspection began, the inspector would fill out in detail what is called the "Safety and Health Report," or the C.A. 15, on which he described the injury record of the plant, its safety program and its effectiveness, the number of employees, a general description of the plant, and a list of the violations found. The LSB never sent this completed form to the company or to any other party, but entombed it in the bureau's files. Another form, the "Notice of Safety and Health Violations," or the C.A. 16, simply indicated the name of the company, the date of the safety survey, the violations found, and the date upon which violations must be remedied. The LSB sent this form to the company but refused to release it to other members of the public, including union officials who represent the workers endangered by the hazardous conditions described on the form.

While in some cases the inspector might be willing to discuss his findings in a general way with union officials, the LSB would not give the union a copy of the notice of violations because, in the words of Eugene L. Newman, head of LSB's Contract Safety Division, "This is a contractual relation between the employer and the government." He added that the information in the notice of violations "is confidential as far as we're concerned."[11] He justified his position by noting that the government does not wish to embarrass employers, a policy of "not taking sides." In fact, by maintaining secrecy, the LSB had been taking sides—with the employer and against the workers. The employer received a report of safety and health violations, while the employee was kept in the dark about hazards to his life and limb.

In 1969, three files were made available to the Study Group before the confidentiality rule was invoked. To test the legality of this veil of secrecy, members of the Study Group sought access to these forms under the Freedom of Information Act. As required by the act, the Study Group first sent a written request to David Swankin, then director of the LSB, asking for the C.A. 15 and C.A. 16 inspection and notice-of-violation reports for companies that were not currently under investigation by the bureau. Swankin denied this request in a letter dated July 13, 1969, when Study Group members refused to sign a waiver form requesting them to promise not to disclose the names of companies mentioned in the reports. The waiver idea emanated from Laurence Silberman, then Solicitor of Labor, whose eagerness to protect corporations accused of subjecting workers to the risk of life and limb was consistent with his "hands-off" policy toward irregularities in the United Mine Workers' presidential election battle—a refusal to use the Department of Labor's legal authority to protect the insurgent candidate (who was subsequently murdered) from election law violations committed by the incumbent.[12]

Informal negotiations with the LSB failed to resolve the impasse, and on August 25, 1969, the Study Group made an appeal to the Solicitor's Office, as required by the act. In a burst of bureaucratic obstructionism, typical of the ways in which government officials make a joke out of the Freedom of Information Act, the Solicitor's Office refused to respond, until Study Group members secured a court order compelling the Solicitor to issue a formal refusal in January, 1970. Only then could the Study Group, joined by the United Auto Workers and the Oil, Chemical and Atomic Workers Union, go into court to determine the legality of the Department of Labor's denial of access to the documents in question. The Department's position before the court was that the inspection and notice-of-violation forms were part of investigative files for use in current enforcement actions, and hence exempt from disclosure under the Freedom of Information Act. When each year fifteen hundred new inspections turn up over forty thousand violations, but an average of only twenty-five formal complaints are issued, it was patently absurd for the Department to argue that inspection reports over a year old were ever going to be used for enforcement purposes. (Indeed, one of the major purposes of the Study Group was to document the Department's failure to enforce.) The Federal District Court in Washington, D.C., agreed and upheld the Study Group's request for disclosure in February, 1971, nearly two years after the initial request.[13]

Winning this lawsuit turned out to be but half the battle

and marked the beginning of a "Catch-22" game with the Department of Labor bureaucracy. Originally, the Department had stored copies of the inspection reports in Washington. In the course of the lawsuit, all these records were sent back to the ten regional offices of the Labor Department, a move that precluded any analysis of inspection and enforcement procedures on a national basis. A Department of Labor attorney remarked to a Study Group member over the telephone that this had not been done to prevent access to the reports (although it certainly had that effect), but to clear more space in the Washington office.

Physical dispersal was not the only obstacle looming in the way of access to these files. A Study Group member, who worked for the Alliance for Labor Action's 1971 Summer Project on Job Safety and Health, sought access to Walsh-Healey files in the New York and Philadelphia regional offices. She expected no problems, since shortly after the termination of the lawsuit the Labor Department had circulated to the regional offices a policy memorandum explaining that these files were now available to the public.

Mr. Joseph Barkan of the New York office told the Study Group member that she could not obtain inspection reports at random, but would have to furnish the names of the companies whose files she wanted to see. When she asked him for a list of all the companies of which inspections had been made, she discovered that his office didn't have one. As it happened, no one did. The only record of plants covered by Walsh-Healey in the New York region was a Department of Defense computer print-out that the inspector used to plot geographic clusters of plants in order to arrange inspections. In addition, it turned out that there were only about sixty files in the New York office. Nonetheless, the Study Group member had to ask the Solicitor of Labor's Office in Washington to telephone the New York office to secure access to the files.

When the Study Group member contacted Mr. Joseph Perzella of the Philadelphia office, he stated that he had never heard of the memorandum explaining the implications of the lawsuit. He referred the Study Group member to the regional solicitor, who was aware of the court decision, if not the memo.

He stated that only files marked "closed" would be available for inspection, a complete misreading of the court's decision.

In a July 19, 1971, letter to Mr. Perzella, the Study Group member referred to the Department of Labor's policy memo, which stated that a list of companies was not a prerequisite to inspection. Immediately upon receipt of the letter, Mr. Perzella called to say that this interpretation was erroneous; that under Labor Department regulations, files had to be specifically identified before they could be produced for inspection; that without the name of the company the file could not be identified, and therefore a list of specific companies was necessary. The Study Group member then asked him whether he had such a list. He replied that he didn't, but that even if he did, he couldn't disclose it. (In effect what he was saying was: "You need a list to see the files open to the public. I don't have a list and neither do you; therefore you can't see the files.")

Again, it took a phone call from the Solicitor's Office in Washington to set things straight. But when the Study Group member asked Mr. Paul Tenny, a Labor Department attorney, for a copy of the memo to the regions explaining the availability of the files so that she would know how to deal with the regional offices, he replied: "It was intra-agency and not factual and therefore exempt from disclosure under the Freedom of Information Act."

The Study Group member finally did get to see the files. They contained nothing remotely approximating confidential information or trade secrets, as the government had argued in the lawsuit. According to the Study Group member's report to the Alliance for Labor Action:

> A sampling of inspection reports from the New York and Philadelphia regions disclosed a great similarity in types of hazards. Many machines lacked guards. For example, in the Danville, Virginia, plant of Corning Glass, the blade of a DeWalt cut-off saw extended over the table. The company was requested to provide protection from the blade either by extending the table or providing stops that would prevent the blade passing the end of the table. Inspectors' descriptions of the sanitary and general housekeeping conditions in the plants rivaled scenes of uncollected garbage in the slums, while small items such as fire extinguisher reinspections were

invariably neglected. Workers were often denied the opportunity to protect their eyes, since washing facilities were not available where required. For example, in a 1970 inspection of Lever Brothers in Baltimore, Maryland, there were "no suitable facilities for quick drenching or flushing of the eyes and body whenever the eyes and body of any person may be exposed to injurious corrosive materials."

Noise pollution was a constant source of trouble. The levels . . . were often above the official watered-down Walsh-Healey limit of 90 dBA [decibels]. Lever Brothers, and Texaco in Westville, New Jersey, are two examples. In the Texaco compressor building, decibel readings ranged from 71 in the control room to 102 dBA in the aisle adjacent to the air compressors. Over three shifts, seven operators suffered five hours of exposure, and while protective equipment was provided, its use was not mandatory. In the turbine room at Westville, the range was 71 to 97 dBA, with two operators receiving eight hours direct exposure. Three areas in the Lever Brothers plant in Baltimore violated the noise standards as of a February, 1971, inspection, exposing the operators to possible deafness.

In fiscal 1969, safety engineers in the regional offices of the LSB performed safety and health inspections in only 5 percent of the firms covered by Walsh-Healey. Out of a total of 75,000 firms affected by the act, only 2,929 were inspected. In 95 percent of these establishments, inspectors found safety violations—indeed, a grand total of 33,378 violations. Out of this number, only 34 formal complaints were issued, 32 hearings were held, and 2 firms received the ultimate sanction of blacklisting. In fiscal 1968, the LSB inspected 1,570 firms, found 48,646 violations, issued 28 complaints, and blacklisted 3 firms.[14]

These amazing statistics were compiled with the help of James Miller, Deputy Associate Solicitor for Litigation in the LSB. When asked about the obvious lack of enforcement revealed by these figures, he replied: "The policy of the Department has always been to treat the act as remedial, not punitive. If we treated it as punitive, we would have half the establishments in the country closed."[15]

An LSB regional inspector expressed the same sentiment in an interview with a Study Group member. While discussing the file on the Emco Porcelain Enamel Co., a major Walsh-Healey violator, 97 percent of whose work was under contract with the federal government, he observed: "Sometimes you have to have a heart; it is not our job to put them out of business."[16]

This sort of attitude completely de-toothed Walsh-Healey, as it signaled the total abdication of any rational enforcement of the law. In an exercise of prosecutorial discretion, the Labor Department had just about reached the extreme of no prosecution, using a limp argument based upon the imagined consequences of a policy of total enforcement. The Department refused to face the real issue, which was how to use its discretion to develop enforcement priorities and utilize the sanction of Walsh-Healey to protect workers in the most effective way possible. Instead, as a result of the Department's timid performance, corporations realized that no sanction would ever be invoked against them and had little reason to take the act seriously.

The LSB's enforcement process further explained its statistical record. When the LSB inspector discovered a violation at the worksite, he would tell management about it—either immediately, or at a conference held after the inspection. During the typical conference, the inspector also suggested acceptable methods of correcting violations and attempted to secure an agreement on a date when all corrections must be completed. The LSB instructed its safety engineers that thirty days was usually enough time to allow for most corrections. However, the engineers routinely granted management requests for sixty days. Requests for any period up to ninety days required—and invariably received—approval from an LSB regional director. Requests for completion dates of more than ninety days had to be approved by the LSB's Contracts Safety Division in Washington. As soon as an agreement was reached between the inspector and management, the regional director sent a letter of confirmation to the employer requesting compliance within the agreed period. This period ran from the date of the letter of confirmation, rather than the date of the inspection—a time-lag often spanning up to three weeks and further prolonging the workers' exposure to known hazardous conditions.

One of the files obtained by the Study Group as a result of the Freedom of Information Act suit illustrates this process. In February, 1968, an LSB inspector visited the Emco Porcelain Enamel Co. and recommended 90 days for elimination of violations he found. The regional director at first reduced this period to 30 days, but then granted an additional 60 days when the company put in a request for 90. In June, when the 90-day extension expired, Emco asked for and received 60 more days. Then a follow-up inspection found that out of the sixty-three violations marked by the inspector in February, forty-three were eliminated, twenty-one were in the process of being eliminated, and five new violations had emerged. Thus, after six months, there were still twenty-six existing violations. Sixty days later, the company asserted total compliance, nine months after the initial inspection. This case was never referred to the regional solicitor's office for prosecution despite the fact that LSB directives indicated 90–120 days as the maximum time period allowed for compliance efforts.

The LSB adopted a method of selective follow-up inspections to determine compliance. Twenty-five percent of the violators were supposed to be checked, but in 1968 the actual figure varied from 3 percent in one region to 80 percent in another. These violators who were not reinspected had to convince the LSB safety staff by mail that they had taken corrective action. Regardless of the seriousness of the offense, the violator completely escaped any penalty if he took corrective action.

At this stage, however, many contractors (especially those with experience with the LSB) realized that corrective action was unnecessary. According to LSB directives, the contractor who failed to make corrections within the specified time should have had his file referred to the LSB's regional attorney for legal action. But the field staff blatantly disregarded these directives. Eugene L. Newman, head of the Contract Safety Division, told the Study Group that some regional directors allowed safety inspectors to perform five or six follow-ups before referring the violator to the legal staff. LSB directives indicated that the maximum time lapse between an initial inspection and referral to a regional attorney should never exceed 90–120 days. In fact, in each of the three case files made available to the Study Group in 1969 (with the names of the companies deleted), this time lapse was seven months.

These cases dramatize the sort of conditions workers must endure as a result of the LSB's casual enforcement process. Each of them involved major threats to worker safety. In one of them, the air workmen breathed in a marble plant contained such high concentrations of free silica that at least two men were coughing blood and another was required to curtail heavy physical labor due to lung ailments apparently caused by inhalation of dust over a long period of time. A second case concerned twenty-seven unsafe and unsanitary conditions in a small manufacturing establishment. Among them were unguarded machinery, ungrounded motors and tools, electrical wiring lying in water, exposed electrical contacts, excessive noise levels, unprotected elevator openings, and inadequate ventilation for the removal of flammable vapors. The third case involved twenty-nine serious violations. The worst two were sanitation and dust. The employee bathroom—which also served as a lunchroom and locker area—was described as slippery and filthy ("slimy water covered the floor"). The entire plant was covered with a measurable layer of dust, which was the result of a cotton-shredding operation. According to the industrial hygienist assigned to the case, airborne dust was so thick that it was impossible to see from one side of the plant to the other. "Frankly," he added, "I don't know why they didn't have a dust explosion before we got there." (Considering the seven-month delay between the first inspection and the initiation of legal action, it is also surprising they didn't have a dust explosion *after* the LSB inspectors got there.) The plant employed unskilled minimum-wage laborers, nearly all of whom were black. Management's attitude toward them was described as extremely callous. In the words of one LSB inspector assigned to the case, "They were getting the work for peanuts, so they treated the employees like animals."

In the light of the LSB's reluctance to proceed against violators, it is not surprising that once the bureau did refer a case for legal action, it was seldom prosecuted with any degree of vigor. Even at this late stage of the enforcement process, Department of Labor attorneys still assumed for themselves discretionary authority to delay and/or drop prosecutions based on continuing violations of the law. In fiscal 1968, the LSB referred two hundred cases to these attorneys for legal action; the attorneys issued only twenty-eight formal complaints.[17]

Note that each of these cases was set down for legal action

only after months of fruitless negotiation by LSB personnel. They involved contractors who stubbornly refused to correct safety and health violations. Every case was one in which the LSB had lost patience. The judgment of the bureau was that the time for voluntary compliance had elapsed. But the staff of Laurence Silberman, Solicitor of Labor at the time of the Study Group's investigation, did not share this judgment. In their view, a referral from an LSB regional director signified that it was once again time to negotiate.

The Solicitor's staff aimed for compliance, not prosecution. They were satisfied with correction of violations and a promise not to break the law again. They moved very slowly. In the three cases made available to the Study Group in 1969, six, twelve, and fourteen months passed before a formal complaint was issued.

In discussions with Department of Labor officials, members of the Study Group repeatedly raised the argument that blacklisting a greater number of firms would have a positive impact on health and safety conditions in plants covered by Walsh-Healey. These officials all agreed that blacklisting only a few more firms would have a substantial "ripple" effect throughout an industry. All agreed that this could be done without increased government expense. Yet the Study Group was told repeatedly, without adequate explanation, that blacklisting was not compatible with high-level Department policy.

Blacklisting would be a costly penalty for many employers. It is not insurable. David Swankin, former director of LSB, has said that "the basic element of employer psychology is cost. The whole thing is how much it costs him. You're in a new ballpark when you're costing him money." The Walsh-Healey Act furnished the government with a ticket to this ballpark, but for unknown reasons the Department of Labor refused to leave the sandlots.

"Mission Safety-70" and "Zero in on Safety"

On October 23, 1970, President Richard M. Nixon announced the inception of "Zero in on Safety," a two-year plan designed to reduce the frequency of injuries to government employees and patterned after the "Zero In" program developed for private industry by the National Safety Council. The new federal program was to make extensive use of National Safety Council promotional materials, which included the council's *Zero In Program Planning Guide*.[18]

At that very moment, in the thirteen miles of underground steam tunnels in the District of Columbia, twenty-five federal employees working for the General Services Administration (GSA) were keeping the federal government's heating system in operating order. "Abandon hope all ye who enter here," warned those who approached Dante's Inferno. No such caption marks the entrances to the tunnels, where workers over the past decade have been facing their own peculiar hell.

> First Week:
> Introduce the campaign without any reference to safety at all. Just use the term ZERO IN as a teaser. This will give the project an air of mystery and arouse interest in the work force. Carry the teaser for a week. ZERO IN buttons can be distributed at this time. . . .
> —*Zero In Program Planning Guide*

Exposed to temperatures that average between 135 and 150 degrees, the workers in the tunnels cannot wear rings or watches or other metal objects (including buttons) because they would overheat quickly and burn the skin. Safety glasses with metal frames, though issued, cannot be worn for the same reason.

There is little mystery about the tunnels. Prospective employees receive a tour before being offered a job, so that they will know what they are getting into. Apparently, this is done on the theory that it is not necessary to protect the health of employees who know about the health hazards to which they are being exposed. This is a reincarnation of the common-law doctrine of assumption of risk, but in a much more pernicious form.

> Second Week:
> During the second week of the campaign the chief executive should issue a statement to all employees emphasizing the organization's dedication to safety and that support of the ZERO IN effort will help achieve an accident-free workplace. It is important to establish management's support of ZERO IN early in the program.
> —*Zero In Program Planning Guide*

Upon hearing complaints about the effects of 170 degree heat on the human body, John Galuardi, GSA's regional administrator, remarked, "Well, some people pay to visit a sauna, don't they?"

Federal employees do not have to pay to work close to ten hours out of a forty-hour week in their cramped, dimly lit sauna. In fact, they even receive a hazardous-duty pay differential of seventeen extra cents an hour for time spent in the tunnels. By some strange economic logic this paltry sum is supposed to justify the federal government's exposing men in good physical condition to the risk of heart attacks and respiratory diseases.

Robert Vaughn of the Public Interest Research Group accompanied several workers on a nighttime drive over the tunnel route. He likened his odyssey to a tour of a battlefield, as the workers pointed out where one employee had a heart attack, where another fell from a ladder, and where another blacked out from the extreme temperature changes. In the winter, the temperature differential between the tunnels and above ground may be well over 100 degrees, yet the men (like Bolivian tin miners, who work under similar conditions) have no facilities to cool them off gradually before they leave the tunnels. Indeed, some workers prefer the summer, because Washington's oppressively humid weather seems cool to them as they emerge from their sauna.

> Third Week:
> Score keeping should start with the third week by pasting the ZERO IN stickers in the daily squares of the scoreboard. . . . Additionally, supervisors should give attention to equipment and material damage.
> —*Zero In Program Planning Guide*

In response to an inquiry from Representative Dominick Daniels (D.—N.J.) in June, 1971, regarding work conditions in the steam tunnels, Mr. Galuardi replied, "We are not aware of any inoperative fan due to bad repair."[19]

One reason for this ignorance was the failure of the GSA and the Labor Department, which had also been notified, to talk with any of the workers in the tunnels. The workers reported that the fan (as well as the telephone) at one location had been out of order for three months, and that at another location two fans were not working.

On November 3, 1971, Delegate Walter E. Fauntroy of the District of Columbia toured the steam tunnels. He found numerous burned-out lights and one ventilation fan not operating. He noted that water was leaking into the tunnel, and that at one point the walkway was covered with mud.[20]

> Fourth Week:
> Sometime during the fourth week, consider the possibilities of local newspaper publicity on the campaign. . . . If you have a public relations or community relations department, it can carry the ball. . . . Such publicity is good for the company and pleasing to the employees who are a part of it.
> —*Zero In Program Planning Guide*

Ever since 1959 workers have been complaining about the steam tunnels to government and union officials. Their efforts have gone unrewarded. Tunnel conditions described in a 1963 GSA report—"sulphuretted hydrogen, a gas generated by sewer waste, which, when breathed, can cause acute poisoning, disorganization of the blood, burning pains in the eyes, nose, and throat, gastritis, headache, convulsions, and paralysis"[21]—still exist today. In 1971, workers in these same tunnels reported the presence of sewer gas, and carbon monoxide as well.

On January 5, 1972, Delegate Fauntroy held a press conference to deplore conditions in the tunnels. Four workers participated. Two of these men have since been transferred from tunnel work and face the possibility of being assigned to lower-paying jobs.[22]

When the work environment in the tunnels was described to him, Dr. Austin Henschel of the National Institute of Occupational Safety and Health (successor to BOSH) observed that 170 degree temperatures were "horribly high," and that the ability to perform hard physical work would be limited at 150 degree heat. He also stated that protective, cooling clothing should be required (the GSA requires only long-sleeved shirts and pants), and that even with such protection, workers should spend only ten or fifteen minutes per hour in the tunnels. The worker should spend the rest of the hour in a cool cubicle or room and should not be permitted to go right up into the street.[23]

The gap between the rhetoric of "Zero In" and the reality of the steam tunnels is characteristic of the federal government's attitude toward the occupational environment of federal employees. The record of the federal safety program that preceded "Zero in on Safety" provides a complete, meaningful picture of how Washington has responded to the problem of job hazards encountered by government workers.

On February 16, 1965, President Lyndon B. Johnson announced "a new, practical safety effort designed to reduce federal work injuries and costs" 30 percent by the end of 1970.[24] This was the birth of "Mission Safety-70" (hereinafter referred to as MS-70), amid great fanfare and a burst of fervor emanating directly from the White House. Shortly afterward, federal agency personnel began to sport MS-70 buttons, and a plethora of promotional materials displaying the MS-70 seal circulated through the federal bureaucracy.

When President Johnson inaugurated MS-70, the injury frequency rate for federal employees, on the basis of data collected by the Bureau of Employee's Compensation, was 7.9 disabling injuries per million man-hours of work. The announced goal was to reduce that figure by 30 percent in five years, so that by 1970 the rate would be 5.5. Occupational diseases were virtually ignored.

At the beginning of MS-70, there was a two-year lag in the reporting of federal work injuries. This meant that the rate calculated in 1965 actually covered 1963. As the program progressed this lag was eliminated, so that by the end of 1970 it was possible to use 1970 injuries in the computations. At the same time, the base year was changed to 1964.

It didn't help much. The frequency rate for 1970 was calculated at 6.6, only 14.3 percent lower than the 1964 rate. Nor was this a steady decline, for in 1967 the rate increased to 7.1 after dropping to 6.9 in 1966.

Some grasp of the scope of the federal job safety and health problem is necessary for an appreciation of the difficulties faced by MS-70. In 1965 the federal government employed 2,586,695 people in a wide variety of jobs. Occupational injuries range from dog-bites suffered by mailmen to the killing of a U. S. Agency for International Development police adviser by urban guerrillas in Uruguay. The Secretary of Labor described the gamut of job hazards in his 1969 MS-70 Report:

> The extensive use of trucks and other vehicles by the Post Office; work with live viruses at the National Institutes of Health; heavy construction operations by the Tennessee Valley Authority and Interior Department; handling of radioactive materials by the Atomic Energy Commission, and of explosives by employees of the armed forces; the printing presses and bindery machines of the Government Printing Office; and, most dramatically and recently, NASA sending three astronauts to the moon.[25]

Almost fifty years of experience preceded MS-70. In 1916 Congress passed the Federal Employees Compensation Act, providing for compensation for federal employees injured while on the job. Though individual departments and agencies began to run their own safety programs, it was not until 1938 that an interdepartmental committee was formed to study federal safety problems and develop a uniform federal safety policy. In 1939 President Roosevelt gave official, independent status to the Federal Interdepartmental Safety Council. But Congressional support never went beyond urging federal agencies to implement safety programs to reduce injuries and prevent accidents. In 1950 President Truman, by executive order, replaced the Federal Interdepartmental Safety Council with the Federal Safety Council, a body composed of representatives of all government departments and agencies, and charged with advising the Secretary of Labor. The Federal Safety Council (FSC) had a small permanent staff. Its members served on a collateral duty basis, whch meant that they did their safety work in addition to their full-time jobs. It tried to promote vigorous safety programs but failed because of a lack of funds, little interagency support, and the absence of a clear definition of purpose and allocation of authority on the part of either Congress or the President.

In 1961, President Kennedy issued a mild statement asking the heads of departments within the federal government "to exert leadership in the establishment of vigorous accident prevention programs to achieve safe conditions of employment."[26] But once again, commitment stopped at the point of verbal expressions of interest. In 1963 the FSC tried unsuccessfully to secure passage of a Federal Employees Safety Act, which would have given the Secretary of Labor authority to develop safety standards, promulgate an accident prevention program, review agency safety plans, collect accident data, and study accident causation. The proposal failed because of interdepartmental jealousies. As John McCart, operations director of the Government Employees' Council of the AFL-CIO, and a supporter of the bill, explained to a member of the Study Group, the agencies resist inroads on their authority and didn't want to see the Secretary of Labor have the power to overrule other cabinet heads. Fred Bishoff, director of safety and health for the Post Office Department, confirmed this view: "An agency head must have total responsibility for running his own program. The Department of Labor should not intervene in the operations of other agencies. The FSC must play a limited role."[27]

The autonomy of the agency heads suggests another weakness of the federal program. No sanctions are available to punish those directly responsible for the maintenance of unsafe or unhealthy working conditions. It is not even possible to assess the guilty agency with fines to be paid out of its operating budget. Without sanctions, federal employees must for the most part rely on the good will of their superiors, a system that does not provide the best of incentives for job safety and health.

Without clearly defined legislative authority or adequate funding, the FSC has been able to play no more than a limited role. With the dawn of MS-70, one would have expected a substantial upgrading of the council. But in fact, the FSC received no increase in funds, and its staff remained the same size for the five-year span of MS-70—three officials, plus secretaries.

All injury statistics during the MS-70 campaign were based upon reports issued by the Bureau of Employees Compensation (BEC), which used as its standard the definition of a compensable injury under the Federal Employees Compensation Act. The agencies kept their own figures, which varied considerably from the BEC's. In 1968, the Secretary of Labor directed all federal agencies to switch to the so-called Z16.1 method of reporting injuries. (In a subsequent chapter, we shall discuss at some length the gross underreporting that characterizes the Z16.1 standard.) The two principal differences between the Z16.1 and the BEC standards were (1) Z16.1 excludes off-premises injuries (such as parking lot accidents) unless they occurred "while the employee was performing the duties of his employment or was under the direction of a supervisor,"[28] whereas such injuries might be compensable, and hence recordable, under the Federal Employees Compensation Act; and (2) under Z16.1, back injuries were excluded if there was no clear record of the accident, or if the employer's physician decided that the work activity did not involve an overexertion that caused the injury,[29] whereas the BEC might determine that such an injury was compensable.

The change-over to Z16.1 did not affect the outcome of MS-70, but it could help the government make future safety programs seem more successful than they actually are. More significant, however, is the unmistakable fact that the BEC is being cut out of federal safety efforts. By law every injury to a federal employee must be reported to the BEC. This has enabled the bureau to compile detailed statistics for each agency, even including the incidence of nondisabling injuries. These efforts have been costly, but one would expect that federal officials concerned with job safety would want the best possible figures upon which to base their programs. A recent economy measure, however, has forced the BEC to reduce its annual report to a scant three-page summary, another example of reality diverging from rhetoric.

There are two other significant shortcomings of the federal job safety effort. MS-70 virtually ignored occupational health and didn't even undertake a sorely needed survey of health hazards encountered by federal employees. We have already described the steam tunnels. Another illustration of such hazards is the exposure to gases found in the State Department in 1969.[30] Duplicating machine operators were breathing in three times the permissible limit of methyl alcohol fumes, which can adversely affect the central nervous system and the eyes. There were no exhaust hoods to carry fumes out of the room. Federal employees may be suffering from a host of exposures to heat, noise, gases, and fumes, as well as work stresses that can precipitate heart attacks, but one would never know it from perusing MS-70 literature.

Secondly, federal agencies pay what is euphemistically called an "environmental differential for hazardous duty," something rather like combat pay. These salary differentials result from an ultimate frustration arising from the absence of sanctions against those responsible for unsafe or unhealthy job conditions. A representative of the American Federation of Government Employees, the union that represents federal workers, told a member of the Study Group that the development of an increasing number of categories for pay differentials is the only real way to put pressure on government agencies to eliminate job hazards. The U.S. Civil Service Commission states the objective of the environmental differential as follows:

> Each agency should have as its objective the elimination or reduction to the lowest level possible of all hazards, physical hardships, and working conditions of an unusual nature. When such agency action does not overcome the unusual nature of the hazard, physical hardship, or working condition, an environmental differential is warranted. Even though an environmental differential is authorized, there is an agency responsibility to initiate continuing positive action to eliminate danger and risk which contribute to or cause the hazard, physical hardship, or working condition of an unusual nature. The existence of environmental differentials is not intended to condone work practices which circumvent federal safety laws, rules, and regulations.[31]

But the very existence of the differential, and the eagerness of the union to expand the number of differentials, suggest that federal agencies and departments have not done all they could to eliminate hazards, which is understandable in view of the absence of effective penalties against safety violators.

Although MS-70 fell far short of its stated goal, the program has made an important contribution to the movement for job safety and health. It clearly demonstrated the futility of hortatory hoopla, unaccompanied by resource commitment, legal sanctions, or the involvement of the workers it was designed to protect. The only expenses (except for the hidden costs of the man-hours spent at interdepartmental and departmental meetings) involved the salaries of three staff people and their secretaries. MS-70 never managed to transcend the superficiality of a paper project, a totally unprofessional gesture, a "farce," as one Labor

Department staffer put it in a conversation with a member of the Study Group. Word issued forth from the President to his agency and department chiefs to improve job safety, and it was expected that somehow this thunderbolt from Olympus would cut injuries among federal employees.

The foregoing description demonstrates that federal regulation in and of itself may not be the optimum response to the job accident and disease problem. The history of federal programs up to 1968 suggests that workers would be ill-advised to place all responsibility for their health and safety in the hands of federal officials. But past failures do not necessarily dictate a minimization of Washington's role. The challenge is to devise ways to take full advantage of federal funds and initiative without becoming enmeshed in the suffocating coils of bureaucracy. The extent to which the Occupational Safety and Health Act of 1970 meets this challenge will be discussed at length in Chapters 8 and 9.

6

After the Coffee Break

The person most intimately affected by the work environment is the worker himself. He must suffer the heat, noise, dust, and fumes of the workplace, eight hours a day, five days a week, for most of his adult life. He must submit to constant exposures that take an invisible, incalculable toll upon his health. His is the daily risk of death or serious injury. One would therefore expect to find the labor unions, organizations upon which masses of individual workers rely to represent them, in the vanguard of the fight to eliminate unnecessary hazards from America's factories and workshops.

In the past, working men and women formed unions to protect themselves against the powerful industries that exploited their labor. Since 1880, mass action and organization have succeeded in enabling the trade union movement to penetrate almost every sector of the American economy. These unions fought for and won better wages, shorter hours, job security, health insurance, and a variety of fringe benefits for their members.

Today between one-fifth and one-fourth of America's workers belong to unions. But this should not obscure the importance of organized labor's duty to help safeguard the total work environment. The unions not only have responsibility for securing safer and healthier work conditions for their own members by means of collective bargaining, but they also have the opportunity to advance the cause of job safety and health for all workers through legislative activity and the fostering of a widespread public awareness of the problem.

Certainly a safe and healthy work environment is as central to the worker's concrete and material interests as are his pay check and retirement plan. A living wage and a decent standard of living are meaningful only if they are coupled with a long and healthy life. It may be true, that as one union official has put it, "You can't eat safety." It is equally true that you may not be able to eat without it.

In 1926, President William Green of the American Federation of Labor wrote, "In the fields of industrial accident and disease prevention, wage earners through their only representative agency, the trade union movement, have an immediate and vital interest and stand ready to help in every possible capacity."[1] This chapter focuses critical attention upon how much the unions actually helped, up until the late 1960's, to reduce work accidents and disease.

Collective bargaining

Through the mechanism of the collective bargaining process, unions can secure for their members a greater degree of protection from occupational hazards than they might otherwise enjoy. Yet the unions have been quite slow to take advantage of this opportunity. Jack Suarez, formerly the health and safety director of the International Union of Electrical, Radio and Machine Workers, AFL-CIO, told a Congressional committee, "In negotiating a contract it appears that safety and health clauses come after coffee breaks."[2] The facts bear out this observation.

Of 560 contracts in the state of New York surveyed by the New York Department of Labor in 1956, 53 percent did not even mention the word "safety."[3] Of the 263 contracts that did contain some provision for safety and health, 210 merely had general clauses to the effect that the company and the union "pledge to maintain safe working conditions," or "pledge to comply with safety laws or regulations." Only 32 contracts specified that "employees may refuse to work on unusually hazardous jobs."

A more recent survey completed by the Bureau of Labor

Standards, tabulating all contracts covering 5,000 or more workers (a total of 252 agreements, covering a total of 4,103,075 workers) shows that 104 (covering 1,705,800 workers) have provisions for safety equipment. The Bureau of Labor Standards conducts separate surveys of contracts for important bargaining issues such as vacation plans and pensions. When Mr. Leon Lorden of the Bureau was asked by a member of the Study Group whether there had ever been a survey of safety clauses, he answered in the negative—a further indication that safety and health clauses are not a prominent part of the collective bargaining process.

In a sample collection of forty-three safety and health clauses compiled by the United Auto Workers (UAW) from its local unions, twenty-four placed upon the company a vague and undefined obligation to provide for the safety and health of employees during hours of work. Ten made no mention at all of any such obligation. Eight clauses bound the company to follow standards set by state or federal regulations, while one bound the company to a list of consensus standards set by private groups. The UAW incorporated this last clause into its Model Contract, despite the fact that these consensus standards are framed by management-dominated groups whose primary goal is to accommodate industry, not protect workers.

The only specific standards the United Steelworkers successfully incorporated into their 1968 contract with the U. S. Steel Corp. for production and maintenance employees were radiation exposure levels set by the Atomic Energy Commission. The Steelworkers' Recommended Safety and Health Program contains a Model Contract that has only one specific standard, the Walsh-Healey limit for permissible noise exposures.[4] (Even this standard permits sound levels higher than many experts recommend.[5])

One subject about which labor unions have usually been willing and able to bargain in the area of health and safety has been the joint labor-management safety committee. When a contract has any provision at all for safety, it will usually refer to the formation of a joint committee. The survey done by the Bureau of Labor Standards on collective bargaining contracts covering five thousand or more workers shows the establishment of safety committees in about 25 percent of the contracts. The study done in New York State shows that of 263 contracts, 73, or more than 25 percent, have provisions for a joint committee.[6] Representatives of international unions continually stress that they encourage their locals to obtain provisions in their contracts for job safety committees. They are quick to point to these committees as evidence of their efforts in the field of occupational safety and health. For example, Anthony P. Gildea, director of safety for the International Union of United Brewery, Flour, Cereal, Soft Drink and Distillery Workers of America, told a House subcommittee in 1968: "Our international union never stops reminding our local unions of the need for establishing machinery for management-labor safety committees through contract."[7]

The safety committees, however, have not turned out to be an effective antidote to occupational hazards. The company and the union evenly divide membership on these committees, but the union has to bargain for the right to select its own representatives on the committee. The contract agreements do not set down voting procedures on safety and health issues, but instead rely upon vague language such as the clause in the United Steelworkers agreement with Kennecott Refining Corp. authorizing the committee to "formulate suggested changes in such [safety and health] practices and rules, and recommend to the Safety Section adoption of new practices and rules relating to safety and health."[8] The refusal of either party to agree will effectively veto any recommendation drafted pursuant to this clause. Some contract language is even less specific. For example, the Aluminum Co. of America's 1968 contract with the United Steelworkers stated merely, "The Safety Committee shall hold periodic meetings."

In addition, safety committees often lack any power of enforcement, and hence become mere discussion groups. The New York survey showed that of the seventy-three contracts specifying the formation of joint labor-management safety committees, only three had the authority to shut down a dangerous work area.

Many union officials have complained about the impotence of their safety committees. Floyd Wilcox, of Local 8–575, Oil, Chemical and Atomic Workers Union (OCAW), described the serious health hazards that result from working with chlorine, bromine, benzol, and dimethysulfate, and then added:

We're in negotiations now, and we have an active safety committee in the plant. But anyone who has served on the committee has the idea that it's just more or less of a formality
to go through; go and look around for tripping hazards and
things like that. They have no real authority.[9]

Anthony Mazzocchi, legislative director of the OCAW, is
also very critical of the functioning of these joint committees. He
told a member of the Study Group that the problem is:

Joint safety committees have no teeth. They cannot shut an
operation down. They will have three men from the union
and three from management, and on important issues where
there is a tie vote, management wins. They are usually
scheduled to meet once a month, but often on the appointed
day, management will call up and say they can't make it.
So the frequency of the meetings starts slipping.[10]

In the same vein, he once told his comrades at an industrial
health conference, "I wish you'd all abolish your safety committees. I know you're very proud of your safety committees, but
they're nonsense."[11]

Yet not all safety committees are "nonsense." It is unfortunately characteristic of the unions to blame all safety committee
failures on management's refusal to cooperate. Yet safety committees, even without enforcement powers, can be an effective
tool to pressure management into taking more vigorous safety
and health measures. Union safety members of Local 6 of the
United Auto Workers continually badger the International
Harvester Co. in Melrose Park, Illinois. Their chairman, Carl
Carlson, convinced the union to purchase a noise monitor for
$350 and to bargain (successfully) for the right to use it in the
plant when the company's monitor was not functioning. This
committee has consistently raised safety and health issues and
has impressed upon the company their importance to the employees. This suggests some value in renovating, rather than abolish-
ing, the concept of the safety committee, with responsibility
being placed upon the unions to appoint members who will be
willing and able to be strong advocates on matters of safety and
health.

Several reasons have been put forward to explain the unions'
failure to negotiate meaningful provisions in contracts to protect

health and safety in the work environment. First of all, except for things like loose wires and rotten scaffolds, present-day safety and health hazards involve complicated equipment and new substances that take a great deal of technical and medical competence to understand. Adolf Wismack, president of an OCAW local, told a House subcommittee:

> I would say where we understand what the problem is we have some success [in negotiations]—you know, a helmet, so you don't bump your head. The obvious hazards we are aware of, but we didn't understand the asbestos hazard until we had listened here this morning, so we couldn't negotiate on that. . . . Some locals are just not aware, are not in possession of the facts to bargain intelligently at all times on the [health and safety] question.[12]

It is certainly true, especially in the realm of occupational diseases, that unions may have insufficient information to bargain intelligently on work environment issues. With so many new chemicals in use by industry, no one can yet know which ones have long-range debilitating effects, and what those effects will be. This does not, however, excuse the unions' failure to make full use of research that has already been done to bargain for standards and controls for chemicals already known to be harmful. This does not excuse the unions for not loudly insisting that research be done to establish safe exposure limits for untested chemicals. Nor does it excuse the unions' own failure to employ doctors who are sensitive to the health and safety ramifications of employee complaints. Locals hire lawyers; they need to engage doctors and scientists as well. In lieu of concentrating their competitive efforts on economics, production, and time studies, the unions could take the initiative away from the companies in the area of health and safety. Yet the UAW employs only one industrial hygienist and one safety engineer; the OCAW has one chemical engineer; and the United Steelworkers utilizes the services of four safety and health experts.

Another common excuse for labor's failure to obtain health and safety standards in written agreements is that it is very hard to get management to agree. Many union leaders claim that management refuses to negotiate on health and safety issues because they view this as an infringement upon their prerogative to operate plants as they see fit. For example, union safety official Arthur P. Gildea has testified:

> There has been continual objection by management in negotiations that negotiation in collective bargaining should not be required to deal with safety clauses. Many unions have been unable to get, although they have recommended various safety clauses to be incorporated in the contract, they have not been successful in being able to negotiate these and get the agreement of management . . . management does not believe that unions should have the right to negotiate safety clauses in the contract.[13]

If this is so, it still does not explain why labor has not fought vigorously against management's attitude. There was a time when management refused to discuss wage rates with labor representatives, claiming that wage schedules were their exclusive prerogative. American working men and women fought long and hard battles to win their right to bargain collectively with their employers for a decent, living wage. There is nothing to prevent them from fighting for the right to bargain collectively for a decent and healthy work environment.

Unions have called very few official strikes over safety issues, but there have been many cases of wildcat walkouts to protest hazardous conditions. Frank Martino, legislative director of the International Chemical Workers Union, told a member of the Study Group that when his men have walked off their jobs to protest a particular hazard, they have usually been successful in getting it remedied. The bargaining power of labor unions derives from their ability to shut down a plant or industry. This power could be utilized to secure more than simply the traditional bread-and-butter benefits. As Frank Burke, safety and health director of the United Steelworkers, once stated:

> Now, what do you expect us to do? Strike? My answer is, "Yes." This industry is going to be confronted the same as the coal industry was confronted. . . . The men refused to dig coal and they walked out of the mines until the condition was rectified. We do not have those privileges and we will never get them until we have the right to strike for them.[14]

A third reason unions have failed to bargain about health and safety involves the cost. Health and safety preventive measures cost money. Companies claim that to succumb to union pressures on safety and health would hurt their competitive position within the industry. They argue that if they spent more on safety, these expenditures, plus the amounts they were spending on wages and fringe benefits, might force them out of business. Thus, management has created a choice between safety and "bread-and-butter." As one union leader put it, "Many employers treat job health and safety as items to be negotiated, to trade off and compromise on. . . ."

It is unfortunate that the unions have allowed management to distort the issue in terms of a safety-versus-bread-and-butter choice, a vicious dilemma that hard-hitting advocacy by the unions could effectively counter. Management should be forced to present data supporting safety-cost claims and to compare it with the cost of work accidents to the company—workmen's compensation payments, property damage, time lost from disruptions caused by accidents, and the cost of training workers to replace experienced employees. In addition, the unions should emphasize the costs to be borne by employees injured or stricken with occupational illnesses: for example, economic losses not covered by workmen's compensation, pain and suffering, and the gradual destruction of health often not covered by the compensation statutes. These costs have nothing at all to do with bread-and-butter.

The crucial point is that bargaining for safety and health affects the physical well-being of workers, and the men and women who put their bodies on the line should be given as much information as possible in order to make an informed, intelligent decision about the safeguards they should seek to include in collective bargaining agreements. The unions have an obligation to provide this information and to question management's unsubstantiated claims about health and safety costs.

Many union leaders have been cautious about raising health and safety issues in such a way as to create conflicts with management. One union safety director, the late Victor Whitehouse of the International Brotherhood of Electrical Workers, told a Study Group member that through his work on various committees of the National Safety Council, "labor and management had started to come together on the issue of safety, until Ralph

Nader stepped in and drove a wedge between them." Another union safety director, Alan Burch of the Operating Engineers (and now a member of the federal Occupational Safety and Health Review Commission), disclosed to a member of the Study Group, "In matters of safety, labor's and management's interests are parallel."

These attitudes reflect either a fundamental misconception about the nature of the problem or else a basic unwillingness to try to solve it. If it were truly in management's interests to maintain safe and healthy work environments, then effective action would have been taken long ago. It may be true that companies are interested in reducing casualty rates that cause lost time and employee turnover. The more serious problems, those caused by fumes, noise, dust, heat, and toxic chemicals are, however, expensive to monitor and control. In these matters the crucial question is, who is going to pay the cost? The worker has been paying it out of his own hide. If management is going to be forced to pay its share, then the labor unions are going to have to put up a fight.

Record-Keeping

Before a union can address itself to the health and safety problems of its members, it must be aware of the scope and nature of these problems. One way to do this is to keep track of the injuries suffered by their members. Another is to record what kinds of diseases their men and women are contracting, and what toxic materials they are exposed to.

Very few unions even attempt to keep these kinds of records. Of forty unions surveyed by the Study Group in the summers of 1969 and 1970, only seven kept any statistics at all on the accident rates, death rates, or disease frequencies of their members. And of these seven, only four were trying to keep records of occupational diseases contracted by their members: the International Association of Asbestos Workers, the Operating Engineers, the International Longshoremen's and Warehousemen's Union, and the Oil, Chemical and Atomic Workers Union.

The organizational structure of many unions makes record-keeping inherently difficult. The officers of industrial union locals work only part-time for the union and have no personnel to keep work-absence records or to correlate absences with particular illnesses. Trade unions that maintain hiring halls could obtain records of individual illnesses, but they would have to secure the services of experts to analyze this data. One possibility is for unions, through collective bargaining, to try to obtain access to their members' health insurance records, which are kept by the companies. Craft unions such as the International Printing Pressmen and Assistants Union of North America have kept their own records of death benefits paid out to the families of deceased members, and these are only now being culled for possible evidence of health hazards.

Dr. Thomas F. Mancuso spoke on the subject of union record-keeping at a conference sponsored by the United Steelworkers of America:

> The health hazards of the various industries, occupations, and trades constitute a large area for investigation and concern. Not only is there comparatively little known about the direct causative relationship of these chemicals, but virtually nothing is known of the accumulative effect and the synergistic effects of sequential exposure to toxic chemicals, dusts, fumes, mists, and gases over the lifetime of the employment of the individual. . . .
>
> Unfortunately, other than sporadic individual efforts by a few investigators, for all intents and purposes, studies of the various international unions, trades, and industries and their needs have not been undertaken. It is only through such studies that the particular health problems of the respective international unions and industries can be detected and recognized.[15]

Of course, expenditure of union resources to collect statistics already gathered by and in the possession of the company should be a last resort. Unions should mightily resist company policy that denies employees access to their own medical records, even when requested by the employee's personal physician. A Johns-Manville plant in Massachusetts, which produces asbestos wallboard, takes annual X rays of its workers to test for exposure to asbestos but consistently refuses to release the results to the workers.[16] The company thus withholds from an individual his own medical history, yet the union has failed to mount a legal challenge to what amounts to a gross distortion of the doctor-patient privilege.

One problem with medical records kept by management is that most workers are extremely skeptical of the diagnoses made from them. The individual worker's view of the company doctor is negative, to say the least. The functions performed by plant doctors range from preemployment screening to determining whether an employee is malingering. They control medical data in a plant and thereby exercise important control over the health of the company's employees.

Anthony Mazzocchi, legislative director of the OCAW, relates that when he was president of an OCAW local at a Helena Rubenstein plant on Long Island, "The company doctor was paid on the basis of how he kept the compensation rates down. I don't know how widespread that practice is. You'd have to do a survey. But I know that in my plant that was how it was done."[17]

Workers are especially concerned about the way company doctors handle occupational disease claims. One employee exposed to a cloud of phenol vapor recounted his trials with the medical profession:

> All I thought I needed was some fresh air. Being near quitting time I went to change clothes and get some fresh air. I didn't get any better but I thought the air coming in the car window would clear my head. But I kept getting worse and by 6:00 P.M. same day I couldn't stand up, headache, dizzy. Called family doctor. He said go to bed and he would call for a prescription. All I could tell him [was] that I was in phenol. The next day my dizziness was gone, but I was shaking badly, pain in base of back, head, pain in kidney area.
>
> Appointment with Dr. _____. He stated that this phenol would kill me and I should stay away from it. He told me to see Dr. _____ [the company director of medicine] and he would fix everything for me. I saw Dr. _____ and gave him a blood sample the next day. A few days later Dr. _____ called me in and said they checked the reactor (where phenol was mixed) and that my doctor didn't know what he was talking about. He said the accident had nothing to do with my condition. . . . Since Dr. _____ called my doctor he refuses to commit himself on anything.[18]

Ray Davidson of the OCAW, in his recent book, *Peril on the Job,* entitled his chapter on company doctors "The Veterinarians," and recounts similar stories of misdiagnoses later corrected by private physicians.

In a paper delivered at a symposium on "Workers and the Environment" held at the annual meeting of the American Association for the Advancement of Science in Philadelphia, on December 26, 1971, Dr. Sidney Wolfe of the Health Research Group gave two examples to illustrate his point that, "What's good for General Motors is good for General Motors doctors":

> A worker hospitalized with mercury poisoning was asked if the company doctor had stopped to see him in the hospital. He grimly replied, "Yes," and added that the company doctor had angrily said, " "What are you trying to do, shut the plant down?" " Another company doctor in a second plant was asked why all workers in the mercury cell area had elevated urine mercury levels. He replied that they might be stealing mercury and taking it home.

The company doctor has an inherent conflict in his job. Any occupationally related diagnosis he makes costs his employer money. In addition, in any litigation arising out of his diagnosis he might be summoned as a witness against his employer.

In a speech to the United Mine Workers convention in Denver in 1968, Dr. Lorin Kerr, assistant to the executive medical officer of the United Mine Workers Welfare and Retirement Fund, described his encounters with company doctors:

> I can vividly remember, twenty years ago when I came with the fund, the constant stream of wheezing, breathless coal miners coming to the area office in Morgantown seeking relief from their struggle to breathe. I can also remember how overwhelmed I felt. Never in my earlier professional experience had I observed or heard of a single industry with so many men who seemed to be disabled by their jobs. I say "seemed to be disabled by their jobs" because doctors said these men rarely had silicosis, and it was unusual to find a physician who suspected that coal dust might be dangerous. The disability was called "miners' asthma," and it was accepted by miners and doctors as part of the job.
>
> In searching for the cause of this shortness of breath, we could not agree with those doctors who claimed it was due to "compensationitis" or nervousness. I can assure you

that were I as breathless as some of the miners I have seen, I would be nervous too. But to claim it was due to nervousness and had nothing to do with the dust was nonsense. I know of one doctor who even maintained that breathlessness was due simply to fear of coal mining.[19]

Ignorance may be a great contributor to the medical misdiagnosis found in the occupational health field because company secrecy over medical records extends to other members of the medical profession. Dr. Hawey A. Wells of the Pulmonary Research Lab in Johnstown, Pennsylvania, spoke out on this point in 1968, before the House Select Subcommittee on Labor:

> Many physicians expressed frustrations at having inadequate medical information to diagnose and treat patients with suspected occupational diseases, much less the ability to prevent them.
>
> Dr. John L. Zalinsky came up to us in Detroit and told of thirty cases of chronic beryllium disease caused by exposure to "safe" levels of beryllium dust [at Brush Beryllium]. He was told by the companies [sic] that if he published this material in the medical literature that he would have to look for another job. He was torn between professional honesty and personal security, and before he resolved this dilemma he died of his second heart attack. His material has never been published.[20]

Dr. Wells also described how the same veil of secrecy may descend over tests used by a company to diagnose serious occupational ailments:

> I have known of a plant that uses manganese, a toxic material that has been known to be dangerous for a long time. Recently, through some bitter experiences, the management of this plant found that it did indeed poison the nervous system causing permanent brain damage in exposed workers.
>
> They have used, for about a year, a simple test, no more complicated than a prick on the finger, that will detect exposure to this material before permanent nerve injury occurs.
>
> How many hundreds of other companies use manganese but don't have the advantage of this simple test for detecting exposure because the test has not been published in this country in the medical literature?
>
> . . . the laboratory that conducts that test has been sworn to secrecy. . . .[21]

The test referred to is a simple examination of the peripheral blood smear for "borophillic stippling." It has been evaluated in the French and Russian medical literature.[22]

Disseminating information

Unions also have an obligation to inform their members about the hidden dangers of the work environment. Contemporary health hazards are often subtle and insidious. If workers are to be able to protect themselves, they must know the danger levels and toxic effects of chemicals and gases with which they work. Locals should be told about available technology, how safety equipment works, what this equipment measures, and who is qualified to operate it. Union newspapers and magazines reach every member and would be perfect vehicles for communicating information about health and safety.

Unfortunately, few unions have made use of these channels. A perusal of random issues of union newspapers and magazines published before 1968 reveals almost no articles at all on the health hazards of the workplace. There were occasional articles on the need for a national comprehensive medical-care plan, but never articles about the specific health and safety problems union members faced in the plants. And when legislative and collective bargaining priorities were listed, the health and safety issue often appeared low on the list.

Since 1968 there have been only a few deviations from this practice of noncoverage. The monthly newspaper and the Washington newsletter of the OCAW have carried articles on health conditions in oil and chemical plants in almost every issue for the past three-and-a-half years. These articles include medical information about the specific chemicals members work with, accounts of accidents that could have been avoided, and information about devices that locals can purchase to monitor the air pollutants in their own plants. The legislative newsletter of the UAW has also carried a few articles on air-monitoring devices in an attempt to encourage locals to learn how to test the atmosphere in their own plants. (This provoked a reprimand from the UAW Safety Department in Detroit, which insisted that management would never allow the use of this equipment.)

Another channel through which the unions might have informed their members about safety and health hazards is the regional and international convention. The idea has been slow to take hold, however. Safety and health issues have rarely been raised at union conventions.

In the past two years the unions have held a few special conferences on the subject of occupational safety and health. The Industrial Union Department of the AFL-CIO took the lead by calling a conference in June, 1969, at which representatives of different international unions and various medical specialists spoke about problems such as byssinosis, asbestosis, noise, and dust. Participants stressed the need for health and safety legislation and urged the international unions to take a more active role in detecting and recording the occupational diseases that are bred in their industries. The United Steelworkers of America held national conferences on the subject in November, 1969, and March, 1971, and the UAW held a conference in February, 1970, for two of its districts. At each of these conferences experts addressed the members about the health hazards they faced. In 1969 and 1970 the OCAW held conferences in every district of the union. These meetings were unique in that the local members were invited to address a panel of scientists and medical doctors about the health problems which they, the workers, perceived they were facing in their own plants. The role of the experts was to answer questions from the members and give specific advice. At the same time, the international officers of the union transcribed the conferences, used them as initial data from which to build a body of information about the problems of their members, and widely distributed copies of the transcripts.

These conferences are a small and late start in the necessary direction of informing workers about their job environment. Employees have a right to know the risks they are facing on their jobs. They should not have to wait for outsiders to inform them. This should be a primary task for the unions.

Research

Most of the medical research in the field of occupational health and safety is yet to be done. Very few of the synthetic chemicals now in use by industry have been tested for their effects on humans. No research at all—not even rat studies—has been done on the combined, or possibly synergistic, effects of heat, noise, gases, and fumes.

The unions have not made sufficient use of the studies that have been done. The federal Bureau of Occupational Safety and Health (BOSH) conducted studies in its laboratories in Cincinnati. But in a survey of thirty unions by the Study Group, only ten replied that they had ever even heard of BOSH. Unions in different industries could have put pressure on BOSH and its predecessors to study the health hazards peculiar to their members. They could have made sure to obtain copies of every study done by BOSH, and to have sent to BOSH samples of the materials their members work with, for chemical analysis.

Another idea that is just beginning to emerge is that unions can sponsor their own research. The United Rubber Workers of America recently signed collective bargaining agreements that require that management shall donate .5 cents per man-hour worked for research to be conducted by the Harvard and North Carolina Schools of Public Health. Each contract provides, however, that a company's failure to act on any findings resulting from the research will not constitute a breach of the agreement.[23] The effort represents an entirely new concept in collective bargaining, and it remains to be seen whether the precedent will be widely followed. The OCAW failed to win adoption, in its 1970 bargaining with the oil industry, of the following provision:

> *Section 6.* The company agrees to contribute $.0010 for each barrel of crude refined to a special Health and Safety Fund, which shall be established primarily for the purpose of research of health and safety hazards and the elimination thereof in the industry. The fund is to be administered by a three-member board, consisting of one representative from the industry, one from the International Union, and the third is to be selected by the industry and union representatives from the medical or science professions and is to be engaged in this type of research.[24]

In its current local negotiations with the chemical industry, the union is attempting to secure the adoption of this clause, with an increase in the company's contribution.

Safety and health staff

The international unions have devoted minimal staff resources to job safety and health. Most international unions have a research department, an educational department, and a legislative department, with full-time staff personnel devoted to particular areas of union interest. But very few unions have a safety and health department, or even full-time people working in that area. Of thirty international unions surveyed by the Study Group in 1970, only seven had any people at all designated as safety staff, only four had anyone designated as health staff, and two had one man who handled both health and safety. One illustration of what these figures signify is that in the UAW, one of the most progressive unions, there were only three men in charge of safety to serve a membership of 1.3 million. And these three men comprise the Compensation and Safety Department, with their work primarily devoted to compensation. (They have since been merged into the Social Security Department.)

The affluent International Brotherhood of Teamsters, with a membership of close to 2 million, had no safety or health staff at the international level. The union's policy was to leave job safety and health to the locals, thus ignoring the information system the International organization could create.

Most international unions have also been slow to use medical doctors and industrial hygienists in a staff or consultant capacity to assist them on safety and health problems. Of the thirty unions surveyed by the Study Group, none had a medical doctor or industrial hygienist on the staff, and only six used medical personnel on a consultant basis. At the 1969 Safety and Health Conference of the Industrial Union Department of the AFL-CIO, Mr. Peter Bommarito, president of the United Rubber Workers, raised the issue of using outside professional help:

> It is unfortunate but true that all of organized labor has been, and is now, at a distinct disadvantage, because it has had none of the technical resources on industrial health and safety that have been continuously available to industry over the years.
>
> Management has, during the last thirty years, created and developed the Industrial Hygiene Foundation to serve the needs of industry under contract arrangements with hun-

dreds of companies; to provide the medical, engineering, laboratory, and toxicological services pertaining to the working environment and specific industrial health problems of a company or industry. . . . Moreover, industrial corporations have found it advisable to employ industrial medical doctors and some technical staff to represent the corporation.

In contrast, there is an absence of professional staff and medical direction in various international unions, although each international union may represent coverage of several hundred individual plants and constitute an accumulated need of responsibility far greater for the international union than for any individual corporation.[25]

Two approaches to this problem came to light at the conference. Mr. Bommarito suggested that all of organized labor join in the establishment of an "Industrial Health Research Institute" to carry on research and provide services for all the international unions. The other approach, put forward by Dr. Irving J. Selikoff and others, is for each international union to establish a relationship with a university or medical school to do research on the problems of their industry. To date, there have been no attempts made to establish a medical institute to serve all the international unions. Several internationals, particularly the Asbestos Workers and the OCAW, have begun working closely with medical schools.

In the past two or three years, several unions have awakened to the problem of in-plant safety and health, and have begun to take active steps to deal with the hazards of the work environment. One approach that several unions have taken is the establishment of a "safety bureaucracy" at the international level. The Operating Engineers and the International Brotherhood of Electrical Workers provide two examples. In both unions, all safety and health activities are left in the hands of the safety department. Alan Burch, former safety director of the Operating Engineers, was a member of seven Presidential commissions on safety and worked actively with the National Safety Council, the American National Standards Institute (ANSI), and other consensus standard-setting institutes. The late Victor Whitehouse, safety director of the International Brotherhood of Electrical Workers, was also an active member of Presidential and National Safety Council committees on safety and health. He also put out pamphlets and films on first aid, which were distributed to the union membership. In both unions, the approach to safety has been from the international level downward. Policies are set at the top and then filtered downward through the membership.

There are certain advantages to the safety-bureaucracy approach. It can serve as a compilation center for statistics on injuries and diseases that are collected by the locals. Officials at the international level have a complete overview of the experience of all the locals, whereby they can locate particular problems and trends. Alan Burch has conducted several surveys and epidemiological studies to try to isolate the occupational diseases to which his members are particularly prone.

On the other hand, safety bureaucrats in the international unions are susceptible to the temptation to engage in "consensus" work with groups whose interests conflict with those of labor. The participation of safety officials such as Alan Burch and the late Victor Whitehouse in activities of the ANSI and the National Safety Council has been of highly dubious value. These are industry-dominated organizations totally responsive to corporate needs. Their work in recent years has not reflected the results of any pressure from labor, and the presence of union officials (particularly specialists in job safety and health) lends an undeserved veneer of respectability to what these groups are doing.

Indeed, even when working on committees set up by the Department of Labor, union representatives have been ineffective. One example is the failure of labor representatives to mount any sort of serious fight in 1969 against the inadequate noise-level standard ultimately recommended by Labor Secretary Shultz's advisory committee on Walsh-Healey Act safety standards. Labor couldn't even muster a majority of public comments in favor of a more stringent noise standard proposed by the Department of Labor in 1968.[26] (Labor's lethargy is apparent from the fact that 85 percent of the 115 statements filed with the Labor Department opposed this standard, and this opposition naturally came from industries most concerned with the cost of complying with stricter noise requirements: mining, textile, and paper companies, and machine tool manufacturers.)

Another problem with safety bureaucracies is their difficulty in responding to complaints from below or outside the structure, especially in the absence of mechanisms at the grass-roots level that can both deal on the spot with these matters and relay their substance to the international office for analysis and possible action.

Letters received by the Study Group in 1970 complained of this lack of responsiveness within the union structure. One correspondent, the victim of welder's disease, wrote:

> There is not a chance the company will ventilate, and they are pressing for us to breathe the fumes 80 per cent of our work time. We will all die if someone, somewhere won't help us. Our union does nothing. The safety committee can't get anything done.[27]

A worker who suspected he had contracted chronic phenol poisoning attempted to find medical help. He wrote:

> The union told me there wasn't a doctor anywhere around who would go against [the company]. I ask them if I had to do as some of the others had done. About nine years ago eight men had to die before anyone would move. At that time the company lied and wouldn't supply information to save these men. They died of leukemia induced from benzene. I asked if the only proof in my case would be in the autopsy and they [the union] said they knew of no doctor who would do anything before.[28]

A safety bureaucracy located at the international headquarters, without any local presence, cannot deal with these situations. The solution is for the unions to develop effective complaint mechanisms at the local level and insure two-way communication between this grass-roots activity and the safety bureaucracy.

In addition, to put the entire health and safety effort in the hands of a few people in the international office will not solve the problem, since the crucial struggle must be waged on the local level, where the workers themselves should participate in the setting and enforcing of standards.

The OCAW has been very active in its pursuit of a grass-roots approach to occupational safety and health. A brief review of some of the activities of the union over the past two-and-a-half years seems in order.

At its 1967 convention, the OCAW decided to make safety and health a top-priority issue and to concentrate major energies and resources in that area. Not long after the convention, the union began holding conferences in every district on the subject, "Hazards in the Industrial Environment." They enabled the OCAW to find out what health problems were bothering their members and at the same time to disseminate useful medical information to the men in every plant. The union has followed up these conferences by distributing questionnaires to every local and encouraging members to write or telephone whenever they have further health problems.

Union publications keep the rank and file informed about the chemicals they are using. Every issue of the *OCAW Union News*, since June, 1969, has contained several articles about specific chemicals or plant conditions that are harmful or dangerous. In addition, the OCAW legislative office in Washington publishes a newsletter that goes to every local. These newsletters also contain information about health and safety problems in the chemical and oil plants.

In 1967, rank-and-file members of the OCAW testified at the Senate hearings on the Natural Gas Pipeline Safety Act, fulfilling a role as worker-citizens in the best sense of the term.

The goal of the OCAW involvement with health and safety has been to arouse an awareness of the problems at the local level, since the men in the plant are the ones affected by environmental conditions and hence should know when and how to act on them. Anthony Mazzocchi, director of the OCAW's Citizenship-Legislative Department, told a member of the Study Group that to focus the issue in any other way would be self-defeating:

> The legislative battle alone is not going to solve the problems at the workplace until the men at the local level are prepared to solve it for themselves. And the establishment of a union bureaucracy to handle health and safety would be counterproductive because it would create the illusion that something was being done—it would absorb the issue before the workers could express what their problems are.[29]

The underlying philosophy of such a program is that as long as the men in the plants have to rely on the companies or outside inspectors, they will never really know whether or not they are safe or healthy. The results of these OCAW activities have proved the value of this approach and are discussed in a subsequent chapter.

The OCAW has also made a substantial financial commitment to its health and safety program. Money has been allocated for the video-taping and transcribing of the safety conferences. Ray Davidson, the editor of the union newspaper, was released from his regular duties for six months to enable him to crisscross the United States and Canada interviewing workers about their job health problems. This material provided the basis for a book, *Peril on the Job* (Public Affairs Press, Washington, D.C.), in which Davidson tells the story of the hazards of the work environment in the men's own words. Recently the OCAW has established a Safety and Health Department in the union's main offices in Denver. A chemical engineer serves as full-time director of safety and health to carry on the work already begun: collecting information from men in the plants and giving them the medical and technical advice they need to remedy their grievances.

Until the late 1960's, the contribution of the trade union movement to the cause of occupational safety and health had fallen far short of the effort labor should and could have expended to protect its constituency against job accidents and diseases. Instead of pursuing comprehensive, high-priority programs to improve the workplace environment, the unions tended to ignore the hazards that make factories deadly traps for workers.

In 1968 the Johnson administration drafted work safety and health legislation that would for the first time heavily involve the federal government in attempts to safeguard workers from job-related perils. The unions found themselves faced with the opportunity to make vital contributions to the development and adoption of what amounted to a fresh approach to job health and safety. They could no longer ignore the challenge, and they had to dive, albeit unprepared, headlong into the fray.

A New Beginning

> It must be our goal to protect every one of America's 75
> million workers while they are on the job.[1]

This simple sentence, tucked away in President Lyndon B.
Johnson's Manpower Message to Congress on January 23, 1968,
gave the immediate impetus to forces that catalyzed an awareness of the extent to which job safety and health remained an
unresolved national problem and ultimately led to the enactment of a far-reaching federal statute regulating the industrial
environment. Broad-based interest in occupational safety and
health did not develop because of any great disaster, or any
startling mass-media exposé of working conditions, or any
pressures from organized labor. It was, instead, the result of a
mildly undramatic concatenation of circumstances.

To begin with, one of President Johnson's speech-writers,
Robert Hardesty, just happened to have a brother working as
information officer at HEW's Bureau of Occupational Safety
and Health (BOSH). Jack Hardesty was a voice crying in the
wilderness for some government action to reduce the toll of
industrial accidents and diseases. He had become appalled at
BOSH's lethargy in fulfilling its responsibilities to the American
working man. With help from a few other like-minded individuals, such as George Taylor of the AFL-CIO and Dr. Hawey
A. Wells, Jr., a pathologist deeply involved in the fight against
coal miners' black lung, he sought to chip away at the wall of

indifference that characterized HEW's attitude toward the physical well-being of the worker. He found another ally in his brother, who began to work at the White House in August, 1965, and thereafter managed to insert an occasional mention of the problem in LBJ's speeches.

Johnson, it should be noted, was not at all averse to speaking out on the job safety and health issue. As early as June 23, 1964, he had sponsored a President's Conference on Occupational Safety and declared that "the question today is not whether we can eliminate the cruel costs of the on-the-job injuries, and disablements, and deaths; it is a question of when."[2] But he had no specific new action program in mind, since he went on to talk of progress "by education, by leadership, by patience and perseverance. . . ."[3]

The first Hardesty-inspired reference occurred on August 18, 1965, in the Rose Garden, when the President swore in John Gardner as Secretary of HEW and observed, "If we could just reduce the sick leave of every American worker by one day a year, we would be adding $10 billion annually to our Gross National Product."[4]

Nine months later, in a speech to the International Labor Press Association on May 23, 1966, President Johnson spoke of the health hazards created by new materials and processes developed by industry since World War II. He urged organized labor to transcend the so-called bread-and-butter issues and "join with us in the effort to improve the total environment."[5] He also called upon Secretary John Gardner of HEW to make an immediate, top-priority study of ways to eliminate existing job hazards and to prevent new hazards from occurring. On September 16, 1966, at the signing of the Federal Metal and Non-Metallic Mine Safety Act, the President again referred to the HEW study.[6]

Despite these forthright expressions of Presidential concern, HEW refused to take its head out of the sand, and the leadership of the labor movement was no less indifferent. The HEW response to LBJ's directive was tepid and proposed no fresh approaches to job health. Neither George Meany, president of the AFL-CIO, nor Walter Reuther, head of the UAW, displayed any real interest in the issue, and their lack of enthusiasm undercut any desire on the part of the White House to press forward with a new program. Nevertheless, in mid-1966 the President set up a special task force on worker security, which was charged with studying unemployment and disability insurance, as well as job safety and health.

Meanwhile, at the Department of Labor, interest in job safety had been sputtering in fits and starts. The fragmentation of various job safety and health responsibilities among the Departments of Labor, HEW, Interior, and other agencies and departments inevitably fostered jurisdictional jealousies and bureaucratic infighting that inhibited the development of a broad federal program to protect every worker. Within the Department of Labor itself, internal bureaucratic wrangling made difficult the assignment of top priority to the promotion of any new approach to job safety. Moreover, as described in Chapter 5 of this book, the mediocre performance of the Department in enforcing the Walsh-Healey Act, the one statute that gave the Department broad authority to regulate a substantial number of workplaces, must have inhibited Department officials sensitive to the occupational safety issue and cognizant of the need for a more extensive federal involvement. Finally, the most pressing occupational-environment problems of the day were those relating to health hazards, an area in which the Labor Department had very little expertise.

But in 1967, the Department began to mobilize its resources in a more sustained effort to obtain new legislation. Reports of a high incidence of lung cancer among uranium miners were the immediate spur. Assistant Secretary of· Labor Esther Peterson visited some of the mines, talked with miners, and was deeply moved by their plight. She pressed the issues before Secretary of Labor W. Willard Wirtz, and, in the summer of 1967, the Department began work on the drafting of legislation that would allow the federal government to regulate work hazards.

At the end of 1967 an occupational safety and health bill was included in a package of legislation sent from the Labor Department to the White House. The Presidential task force on worker security, White House officials, and representatives from other agencies and departments had a hand in shaping the final version of the bill. There was some resistance: the Commerce Department, for example, saw no need for federal action. The U. S. Chamber of Commerce was consulted, and argued that although the federal government should provide financial support to state job safety and health programs, authority to regulate the workplace should remain with the states. Although the proposed legislation would vest broad powers in the Labor Department, the Presidential decision to go ahead with it was not coordinated in any way with the Department. Indeed, Secretary Wirtz later admitted, "We didn't know, frankly, till several days before the Message, that the President had decided to make occupational safety and health a principal element in his program this year."[7]

Nonetheless, shortly after the reference in the Manpower Message, Congressman James O'Hara of Michigan and Senator Ralph Yarborough of Texas introduced the administration's bill before the House and Senate, and the Labor Department took the initiative in marshaling support for the new legislation.

The bill authorized the Secretary of Labor to set and enforce safety and health standards for all workplaces not already regulated by any other federal agency, to inspect workplaces subject to the provisions of the act, and immediately to shut down machinery or processes threatening imminent harm to workers. The Secretary of HEW was directed to conduct research into occupational safety and health hazards and to develop criteria for standards. The bill gave no rights or responsibilities to workers or their unions, nor did it mandate the use of existing industry-wide consensus standards developed by private organizations. (Later versions of the act contained provisions for both worker rights and consensus standards.)

On February 1, 1968, the Select Subcommittee on Labor of the House Committee on Education and Labor began hearings on the bill, and the Subcommittee on Labor of the Senate Committee on Labor and Public Welfare followed suit two weeks later. Labor Secretary Wirtz personally made strong statements in support of the bill. Representatives of HEW reflected the Department's customary lack of enthusiasm at the hearings, and HEW Secretary Wilbur Cohen did not testify.

As expected, business interests voiced their strenuous opposition to any federal regulation of the workplace. The U. S. Chamber of Commerce, the National Association of Manufacturers, the American Iron and Steel Institute, and the Manufacturing Chemists Association led the fight against the bill.

One would have expected these and other like-minded groups to have presented data and analysis that would have purported to prove beyond cavil that the private sector and the states had done an adequate, if not praiseworthy, job in reducing the toll of work accidents and diseases. But such was not the case. Surprisingly, their representatives testified in sweeping generalizations and furnished little documentary evidence. For example, Otto F. Christenson, a consultant to the American Small Business Administration, declared that the "bill would in fact add unnecessary federal spending, encroach upon states' rights, and enable unnecessary interference with industry."[8] He also denounced the idea of turning the Secretary of Labor into a "Czar of Safety." But when asked by Congressman William D. Hathaway about job safety and health progress in Mississippi, a state that had done little to protect workers, he gave the classic reply: "I don't know any more about Mississippi than I know about the moon."[9] Commented Representative Hathaway, "That is the problem."

Opponents of the bill, both on the Congressional subcommittees and among the witnesses, tried mightily to divert attention from the real issues by making much of two irrelevancies: the failure of the drafters of the bill to consult with state officials,[10] and the fact that the Labor Department used outdated photographs of industrial accidents in a brochure prepared for the hearings.[11] Yet representatives of state governments offered little of substance in their testimony, and Assistant Secretary of Labor Esther Peterson introduced into the record newspaper accounts of recent job accidents that were similar to those depicted in the photos.[12]

The private groups and trade associations whose representatives testified at the hearings did enjoy one major success. They shielded industry from the glare of embarrassing publicity. At the 1968 hearings not one single corporate executive testified in person on behalf of his company. Thus, members of the Congressional subcommittees were not able to ask specific questions relating to the activity of any one company.

Industry's response to the proposed legislation went far beyond testimony in Washington. The most vituperative attack on both the administration bill and its supporters came from the U. S. Chamber of Commerce in the April, 1968, issue of *Nation's Business*, the chamber's trade journal. An article entitled "Life or Death for Your Business? Labor Secretary Wants the Power to Shut You Down in the Name of Health and Safety"[13] quoted Labor Secretary Wirtz out of context, implying that he was proposing that federal inspectors be drawn from the great pool of hard-core unemployed and be trained to exercise tyrannical authority over industry. The author described how an unemployable young man (black by implication) whose "main experience consisted of cashing welfare checks," and who had just been turned down for employment at a particular company, returns as a federal inspector. He threatens to ". . . padlock your gates and have you fined one thousand dollars a day if you don't do as he says. The young man—who knows nothing about your business—then tramps through your plant, without a warrant, ordering you to take costly steps to improve 'safety and health.' "

Secretary Wirtz had in fact endorsed no such program, as the following excerpt from the hearings will show:

> CONGRESSMAN SCHEUER: Do you feel that as a source of manpower, both for training workers and inspecting physical premises, that we might look to the hard-core cadre of unemployed at present and develop a designed program for training the presently long-term hard-core unemployed to work as subprofessional aides to professionals in both training and inspection? . . . .
> SECRETARY WIRTZ: . . . what we are talking about here obviously involves the kind of thing on which there could be development of a subprofessional level of competence which would be a great help. So that my answer to your question would be an unqualified affirmative.[14]

The *Nation's Business* article proceeded in its "kick-'em-in-the groin" fashion to declare that the proposed bill ". . . opens the door for the Labor Secretary to: rewrite local building codes; revise local fire regulations; cancel any professional football game should he decide, say, that tag football would be safer and healthier than tackle." The author conjured up the specter of

> featherbedding unions [getting] a Labor Secretary to decide that certain new automated devices and labor-saving materials are unsafe or unhealthy. Once the Labor Secretary starts ordering what the working conditions must be in firms, unions can mass their bargaining powers on the juicier aims of bigger wages and shorter hours.

Reprints of the article were widely distributed to members of Congress and members of local chambers of commerce. The reprints elicited numerous letters to Congressmen from interested parties, echoing the language of the article and demanding that the bill be defeated.

Organized labor hardly lived up to the images created by *Nation's Business*. Union leaders saw no reason to exert themselves on behalf of the administration bill. The labor press refused to respond to the scurrilous attacks by the U. S. Chamber. AFL-CIO President George Meany made a rather perfunctory appearance at the House hearings. Other union officials spoke at both hearings, and were often embarrassed by questions from Republicans inquiring about union job safety and health efforts, especially through the mechanism of collective bargaining.

The events that changed President Johnson's political fortunes in the spring of 1968 all but aborted the occupational safety and health movement in Congress. The normal priority any administration bill enjoys quickly evaporated. It was never reported out of the Senate Committee on Labor and Public Welfare. The House Committee on Education and Labor did adopt a compromise version of the bill, but the Democrats lacked the votes in the House Rules Committee to get the bill to the House floor in 1968.

Miraculously, the impetus supplied by the 1968 hearings kept interest in job safety and health alive. The Democrats, although they lost the White House, were determined to try again. The concurrent campaign for coal mine health and safety proved helpful. Publicity generated by the coal mine explosion that took seventy-eight lives in Farmington, West Virginia, on November 20, 1968, the agitation by miners demanding workmen's compensation for black lung and better safety standards for the mines, Congressional hearings on conditions in the mines, and the passage of the federal Coal Mine Health and Safety Act all served as a constant reminder of the larger, unsolved problem of job safety and health conditions in the United States.

But the most surprising source of apparent support came from the Nixon administration's attempt to grab hold of the safety and health issue. The Republicans were making a loud pitch for blue-collar support and had unleashed a torrent of rhetoric concerning the "silent majority" and the "forgotten Americans." Therefore, they had to take some measures to address blue-collar concerns.

In early 1969, the Democrats once again introduced into the House and Senate job safety and health bills similar to the legislation that had been stalled during the previous year. On August 6, 1969, the Republicans countered, as Congressman William H. Ayres of Ohio and Senator Jacob Javits of New York introduced the Nixon administration's occupational safety and health bill.

Some insight into this legislation was immediately provided by Senator Javits, who issued a statement spelling out reservations he had about the bill that bore his own name.[15] He said he had a difficult time accepting the setting and enforcing of standards by an independent board, whose members were appointed by the President rather than by the Secretary of Labor; the number of exemptions from coverage; and the extensive use of the courts, which would prolong enforcement.

If Senator Javits had reservations, organized labor had fits. Union officials took the position that they would prefer no bill at all to the "abomination" concocted by the Nixon administration. On the other hand, in a drastic turnabout, the same business groups to which any federal law had been anathema one year ago suddenly discovered that they could live with, and even support, the Javits-Ayres bill.

In 1969 and 1970, the House and Senate subcommittees held more hearings on occupational safety and health, and the passage of some legislation seemed inevitable. The controversy now focused on what kind of bill should be adopted.

The entire record of the 1968–70 hearings on the industrial

environment fills six volumes. From this mass of testimony and data it is possible to separate out several key issues and revelations that bore heavily upon the need for federal legislation and the shape such legislation should take.

Worker carelessness: the nut beneath the hard hat

One of the most persistent of the arguments mounted against broad federal involvement in the struggle against work accidents and diseases emerged from the notion that the overwhelming majority of job injuries result from worker carelessness; therefore, the proper and better approach to occupational safety is to educate employees, rather than impose mandatory standards upon employers.

Some companies have gone to great lengths in their efforts to "teach" safety and motivate workers to be careful. Many corporations give Green Stamps to workers with outstanding safety records.[16] The *Wall Street Journal* recently reported:

> Officials of Weyerhaeuser Co.'s Fitchburg, Massachusetts, plant have a practice of calling employees' homes and asking family members who answer to repeat the plant's current safety message. If they get it right, the families win trading stamps.[17]

An upstate New York manufacturing plant has adopted "Operation PIG" (*P*revent *I*njury and *G*rime), whereby the department with the poorest safety record over a given period of time must take care of a live pig (presumably caged) until some other department compiles a worse safety record and thus qualifies for custody of the pig.[18] A department with an accident-free record over a given period of time also wins a pig, but in the form of a roast pork dinner.

In line with the argument about worker carelessness were requirements incorporated into the Javits-Ayres bill for research into worker behavior. The Secretary of HEW was directed to "conduct research into the motivational and behavioral factors relating to the field of occupational safety and health."[19]

The worker-carelessness theory surfaced time and again at the hearings. According to Raymond M. Lyons of the National Association of Manufacturers, "It has been estimated that 75 to

85 percent of all such occurrences [accidents and injuries] have been caused by a negligent or unsafe act on the part of the individual."[20] He later cited with approval the slogan, "safety is sold, not told."

J. Sharp Queener of the U. S. Chamber of Commerce told the House subcommittee, "We find that 80 to 90 percent of the injuries which are occurring in our company [Du Pont] are due to a human failure rather than a piece of equipment, a machine, or so on."[21]

Finally, Charles A. Hagberg, an official with the state of Wisconsin's Department of Industry, Labor and Human Relations, disclosed: "It appears from preliminary reports of [a Wisconsin] study that if one were to define an unsafe condition as one covered by our code, that unsafe conditions account for 7 or 8 percent of the injuries, and unsafe acts constitute the reason for the preponderance of injuries."[22]

The minority report of Republican members of the House Committee on Education and Labor opposed to the 1968 Occupational Safety and Health Act also adopted this line:

> There is no reason to believe that federal controls would materially reduce the rate of such accidents or injuries. Safety authorities have estimated that three-quarters of accidents on the job result from unsafe acts rather than unsafe conditions.[23]

A closer look reveals that the worker-carelessness theory is a hoax. It is a version of the "nut-behind-the-wheel" argument used in the unsuccessful attempt to stop legislation giving the federal government authority to impose performance standards upon automobiles. As hoary as the work safety movement itself, the worker-carelessness argument has a very shaky basis in reality. Although one cannot deny that *some* work accidents are causally related to worker carelessness, this does not mean that they all are. Nor does it mean that the frequency and severity of these accidents cannot be substantially reduced by designing the work environment and work practices to take human failings into account.

Dr. Alice Hamilton, one of the first physicians to work in the field of industrial medicine in the United States, was also one of the first debunkers of the myth of worker carelessness:

In smelting, as in white-lead production, and pottery glazing, and coating iron tubs with enamel, it was the pollution of the air with lead that one had to fight; that and the rooted prejudices of the employer who held firmly to the belief that poisoning was caused by the worker's own carelessness, by handling his food and his chewing tobacco without first washing his hands.[24]

If employers were quick to blame workers for contracting industrial diseases, they were even quicker to fault them for causing occupational accidents. Crystal Eastman, in her classic *Work-Accidents and the Law* (published in 1910), reported the following vignette, which demonstrates how little management attitudes on work safety have changed:

> "So you've come to Pittsburgh to study accidents, have you?" says the superintendent, or the claim agent, or the general manager, as the case may be. "Well, I've been in this business fifteen years, and I can tell you one thing right now—95 per cent of our accidents are due to the carelessness of the man who gets hurt. . . ."[25]

Yet in her research and writing more than six decades ago, Ms. Eastman was able to point to a study of occupational fatalities that showed only 32 percent were caused by the victim's own carelessness, and in over half of these cases, the act or omission of a fellow worker, a foreman, or the employer also contributed to the death.

An analysis of fifty thousand work injuries made some forty years ago revealed that 10 percent were due to mechanical hazards, while 90 percent resulted from· some human failure.[26] But the study also concluded that 88 percent of the injuries caused by human failure could have been prevented by proper supervision on the part of management.

A 1967 study of industrial injuries in Pennsylvania came to the conclusion that about 26 percent of the accidents occurring in that state during the year were directly caused by employee carelessness.[27]

Opponents of federal job safety and health legislation reiterated certain claims about worker carelessness and bandied about vague statistics, but they never presented any thorough, scientific studies of accident causation. One of the authors of this book attempted to find substantiation for Charles Hagberg's claim that preliminary reports of a Wisconsin study showed that unsafe acts account for over 90 percent of the work injuries in the state. He learned that the "data" to which Mr. Hagberg was referring were based upon "fragments of information" collected from a small region in the southeastern corner of the state.[28] (A further follow-up inquiry elicited the following reply from a Wisconsin job safety and health official: "Because of our implication in the new Occupational Safety and Health Act of 1970 we have dropped for this period of time our study on the relationship between industrial accidents and unsafe acts and conditions.") * Upon slender reeds such as this rests the myth of worker carelessness.

In any event, it should be noted with emphasis that these arguments related only to work accidents and did not in any way justify federal neglect of the occupational disease problem.

A variation on the worker-carelessness dodge provided one of the more amusing episodes of the Congressional hearings on occupational safety and health—a dialogue between Senator Jennings Randolph of West Virginia and Ralph Nader in 1968. Senator Randolph invoked the specter of alcoholism among workers as being a prime cause of industrial accidents. Though he cited no factual basis for this charge, he indicated his belief that "there are many accidents which are directly attributable to intoxication in one form or another." In response, Nader suggested that about 3 percent of the nation's working force suffers from alcoholism, but that there were no studies examining the relation-

---

* The Department of Labor subsequently commissioned the Wisconsin Department of Industry, Labor and Human Relations to study the relationship between state safety code violations, as cited by inspectors, and work injuries resulting in workmen's compensation awards. The final report (dated September 28, 1971, and on file at the Labor Department) predictably reflects the Wisconsin officials' blind embrace of the worker-carelessness myth, although they are now claiming that only 45 percent of the accidents they studied resulted from worker behavior. The biased nature of this statistic is evident from the source of their work injury data—forms filled out solely by employers, who have a definite interest in promoting the notion that industrial accidents and diseases are caused by employees themselves. The form's bias is also patent: it asks only whether the employee was intoxicated, failed to use a safety device, or failed to obey rules. There is no mention of unsafe conditions.

ship between worker alcoholism and job accidents. He then cited a survey published in *U. S. News & World Report* showing that 27 percent of America's executives have three or more drinks a day. "Clearly," he concluded, "one of the occupational hazards for top management is the ability to take long lunches and imbibe martinis, and workingmen just don't have martinis during lunch."[30]

Worker carelessness, as well as alcoholism and the more recent problem of drug abuse among workers,[31] requires thoughtful study transcending the self-serving, superficial statements made by opponents of broad federal legislation during the hearings. We need to know the extent to which the dispiriting, dehumanizing nature of the occupational environment forces workers into the escapism of daydreams, drink, and drugs and to devise ways to reshape that environment and attack the problem at its roots.*

The performance of professional groups

The job safety and health hearings attracted the interest of many private professional groups whose members dealt with work accidents and/or diseases. Representatives of safety, engineering, health, and medical groups appeared before the Congressional subcommittees. Virtually all of them took the viewpoint of business and advocated the adoption of the weakest possible legislation.

The performance of the National Safety Council was particularly intriguing, for it clearly demonstrated the degree to which the council responds to the wishes of the corporations that fund it. The council came into being as a result of the First Cooperative Safety Congress held in Wisconsin in 1912. Its first mission was to promote industrial safety, but in 1914 its functions expanded to include automobile, home, and public safety. Since these three areas account for 86 percent of all accidents in the United States, the council naturally devoted most of its attention and resources to them. But the occupational environment was not completely

* For a promising approach to the study of these problems, see the Hearings on Worker Alienation before the Subcommittee on Employment, Manpower, and Poverty of the Senate Committee on Labor and Public Welfare, July 25–26, 1972.

forgotten, as one of the council's objectives remained "to further, encourage, and promote methods and procedures leading to increased safety protection and health among employees and employers." In 1953 the National Safety Council received a federal charter as a nongovernmental, public service organization.

When the federal job safety and health bills were first introduced, the council seemed on the verge of taking seriously its obligation to act in the public interest. In a confidential memorandum dated February 24, 1968, the National Safety Council's Federal Safety Committee strongly endorsed federal standards:

> 1. The federal government has the right and obligation to be concerned with occupational safety and health.
> 2. There is need for reasonably uniform standards dealing with occupational safety and health, and there should be federal legislation designed to achieve this end in the most efficient and effective manner.[32]

The remainder of the memo recommended the formulation of standards based on the principle of industry-wide consensus, and a system of state plans that would enable the states, under the supervision of the Secretary of Labor, to enforce these standards. The basic premise of the memo, acknowledgment of the necessity for a federal role in promulgating and enforcing standards, departed significantly from the position being taken by other private groups.

But within three months, the council had undergone a change of heart about the need for a strong federal role. A new position, approved by the council's Executive Committee and its industry-dominated Board of Directors, made no reference to the need for a federal statute and for uniform, national safety and health standards. The council refused to take an official position on these points. But its view of the federal government's proper role was implicit in the testimony of council President Howard Pyle before the House subcommittee in 1969, when he stated:

> . . . it is our judgment that some of the most effective contributions that government can bring to the occupational safety movement are in the areas of research, dissemination of safety information, and education and training for employers and employees.[33]

Pyle's conception of the government's responsibility, his acceptance of the worker-carelessness myth, his insistence on unenforceable consensus standards, and his emphasis upon restoring as much safety and health responsibility as possible to the states, made it clear that the National Safety Council's first allegiance was to the corporations that funded it, rather than to the men and women who comprised the accident statistics the council published.

The American Society of Safety Engineers (ASSE) took no position on the need for a federal statute but instead downgraded the idea of mandatory regulation and advocated further research into accident causation and frequency.[34] The society was well suited to talk about delay since it had been procrastinating on the subject of job safety since its inception in 1911. Indeed, after sixty years of existence, the ASSE has only recently begun to take its first steps in research and education in the area of industrial safety.

The American Society for Safety Research, a corporation set up only recently by the ASSE to stimulate research projects, conducted a survey of accident research in the United States and found that out of the $23 billion spent for research and development in 1966, only $7.5 million went for accident research. Out of the total of 147 projects funded by this latter sum, only 24 were devoted to industrial accident studies.[35]

In 1969 the ASSE finally got around to surveying the status of safety education in the U.S. Its report lists 1,011 safety courses in college catalogues.[36] Of these, 367 were minimally related to industrial safety, especially the 280 driver-education courses. No attempt was made to ascertain the actual work safety content of all these courses.

Most of the society's members were trained as mechanical, electrical, or civil engineers, and subsequently went into safety work. By the society's own estimate, the demand for safety engineers is so great that for every person in the field there are three jobs available. Yet only recently did the ASSE bestir itself to design a four-year undergraduate curriculum for safety engineers. At the 1969 Senate hearings, A. C. Blackman, the society's director (and presently the assistant director for safety of the

National Institute of Occupational Safety and Health), spoke with some pride about the curriculum:

> MR. BLACKMAN: This is a first step. This is a first effort on our part to set forth a four-year bachelor's degree program at the college level.
> SENATOR WILLIAMS: Where is it offered?
> MR. BLACKMAN: It isn't offered any such place at the moment.
> SENATOR WILLIAMS: Oh.[37]

The American Medical Association (AMA) took a strong stand against the basic thrust of the Johnson administration bill. Dr. R. Lomax Wells, chairman of the AMA's Council on Occupational Health, presented the association's position at the 1968 House hearings:

> On the matter of standards, we believe that the creation and enforcement of mandatory standards for occupational health and safety . . . should not be undertaken by the federal government.
> Occupational health and safety standards have been properly vested in the states, where the standards and the enforcement can be adapted to their particular geography and industrial conditions.[38]

The AMA did favor federal grants to support state programs and to fund research into job safety and health. In 1969, the AMA came out strongly in favor of the Javits-Ayres bill.

The AMA's posture should be judged in light of the medical profession's record in the field of occupational medicine. Although the prevention of most modern industrial diseases would seem within reach—through the elimination of disease-causing materials and processes and the physical protection of workers exposed to potentially hazardous substances and stresses—the practice of preventive medicine in the field of occupational health is virtually nonexistent. According to Dr. George Wilkins, a professor at the Harvard Medical School and a member of the Occupational Health Council of the AMA, only twelve medical schools teach occupational health, and twenty-five to fifty graduates enter the field each year.[39] This reflects the orientation of the medical profession toward the curing rather than the prevention of disease.

The AMA's Council on Occupational Health, in existence since 1937, holds three meetings and a congress annually. Its audience is the physician and nurse within the AMA, and its primary role is to provide elementary education for the medical profession. Not until 1968 did the council initiate a registry of adverse reactions to physical hazards and chemicals in the work environment. But the information obtained is confidential and subject to release only at the council's discretion. Reports are obtained from doctors who volunteer to send them. California is the constant and almost sole source of the two hundred to three hundred cases reported to the council each month, primarily because California law requires occupational diseases to be reported by physicians, and hence it is no problem for doctors to send copies of these reports to the council. This experience suggests that it is incredibly naive to think that a voluntary system will furnish data from which effective conclusions can be drawn. Yet Dr. Henry Howe, a specialist in occupational health from the AMA's Chicago headquarters, testified to a Senate subcommittee, "Our hope is that physicians will report these cases to their professional association when they would not report it to the government." New Jersey's Senator Williams was incredulous: "This has been said and we will be told that fifty years from now. The committee will be hearing the same statement and hearing the same testimony."[40]

The council estimates that 90 percent of all physicians in private practice have contact with occupational health problems. But physicians are not oriented to recognize them as such, and their patients cannot be expected to recognize the causes of these illnesses.

In view of the medical profession's meager contributions to the cause of occupational health, it is not surprising that the AMA could offer Congress no better solution than to let the states maintain their authority over the occupational environment and encourage the nation's doctors to continue, on a purely voluntary basis, their feeble efforts to improve the health of the American worker.

One private group that did not appear at the hearings might have supplied Congress with some insight into the low estate of scientific research into job health problems. The Industrial

Health Foundation, a nonprofit research organization founded in 1935, is industry's answer to the need for scientific exploration into the causes of industrial diseases. Union Carbide, U.S. Steel, and General Motors are among the four hundred corporations participating in the foundation. According to the foundation's 1969 tax return, $360,900.95, a substantial majority of its income, was contributed by member industries sponsoring projects being conducted by the foundation. It provides engineering, legal, and medical services to members and is closely associated with Carnegie-Mellon University in Pittsburgh.

The objectivity of the foundation's research is called into question by its stated aim of warding off those federal agencies that want "to assume czar-type responsibility and authority in industrial health matters."[41] This goal is to be reached by "showing the federal authorities that their intervention is both unnecessary and undermining to the democratic process and sound industrial management." To meet this goal, the foundation undertakes research projects the results of which "will show what shall be removed from the work environment to conserve the health of the work force and should prevent unwarranted exploitation of the employer through false claims by knowing what does and what does not constitute a health hazard."[42]

A brief glance at the research reported at the foundation's annual meeting in 1968[43] reveals how these quasi-ideological directives are carried into action. Hypersensitivity is emphasized: the reaction to a chemical is not so much the fault of working conditions as of the worker's own biological makeup. Disability and sick absences have increased because of outside factors and "changes in the way people feel when ill." Hearing loss is due more to susceptible ears and the natural aging process than to industrial noise. And so it goes.

One private professional group that might have offered testimony to offset the impressions made by the AMA, ASSE, et al. was the American Trial Lawyers (ATL), an association of personal injury lawyers. ATL's workmen's compensation section contains men who handle compensation claims on behalf of injured employees and are uncommonly knowledgeable about conditions in the plants and workshops. Yet neither representatives of the association nor any of its members, as individuals, appeared in Washington to describe from their own experience the realities of the industrial environment or to advocate meaningful legislation to protect the safety and health of the worker.

Workmen's compensation and the occupational environment

The degree to which existing approaches to occupational safety and health furnished economic incentives for the prevention of work accidents and diseases remained unclear throughout the Congressional hearings. Apart from governmental regulation, and limitations imposed by collective bargaining, the principal costs incurred by companies as a result of job hazards were premiums paid for workmen's compensation insurance. Indeed, one objective of workmen's compensation is to prevent accidents by making safety pay. This assumes that compensation insurance costs are sufficiently onerous to make it worthwhile for employers to spend money on safety precautions.

Workmen's compensation insurance rates are set by the National Council of Compensation Insurers (NCCI), an organization created by the insurance industry for that purpose. The NCCI divides businesses into classes according to the nature of their work, and regardless of payroll size, and then calculates a so-called manual rate based on the average accident experience of that class in each state. If this rate is approved by the government agency that regulates the insurance industry in that state, each member of the class pays the same rate per one hundred dollars of payroll, regardless of the individual firm's own accident record.

A firm that pays more than five hundred dollars in total premiums receives an experience rating from the NCCI. The amount of workmen's compensation benefits paid out over the preceding three years as a result of injuries and diseases suffered by employees of the firm is measured against the average in that particular state. If the company's experience was better than average, it receives a premium credit, and if worse, a debit is assessed.

Approximately 15 percent of all employers, paying 85 percent of all workmen's compensation premium dollars, receive an experience rating. This should provide a real safety incentive, but there is uncertainty about its actual impact. As Lee Holmes, counsel for the Washington office of the American Mutual Insurance Alliance testified at the hearings, "I think we feel it has had a definite effect, but I don't know that anybody has ever sought to analyze this effect on an industry."[44]

For the 85 percent of employers not covered by experience ratings, the existing manual rating system has the effect of burying their accident records in the class total and keeping their insurance premiums down until the class rate as a whole rises. As one commentator has explained:

> It is not practical to make experience rating applicable to very small risks. Such employers would be unduly penalized by even one moderately severe accident. This would be contrary to the insurance principle of spreading the risk. Of course, these employers get the benefit of improved experience applicable to the class as a whole.[45]

Smaller firms, therefore, profit from the good experience of larger firms, without any immediate consequences for their own poor records. For example, in Washington, D.C., a worker's death in a painting firm employing two men would not cost a single penny in increased compensation rates unless enough other painters in the district had similar accident experiences to justify a rate increase.

The experience rating system further weakens safety incentives because it links the costs of compensation insurance to the amount of benefits paid out after an accident has occurred, rather than to the level of safety and health protection in a plant. It is impossible, therefore, to relate plant expenditures on safety and health to the cost of compensation premiums.

Finally, occupational diseases constitute the most serious hazard to the American worker today, yet many of these illnesses are not covered by state workmen's compensation acts. In addition, as we have pointed out in Chapter 2, workers may not realize they are suffering from one of these diseases, or that the disease is work-related. Therefore, the experience rating system will scarcely, if at all, reflect the actual incidence or severity of occupational diseases at a particular plant and thus provides virtually no incentive at all for the prevention of job-related diseases. Less than 1 percent of all compensation payments, the House learned, goes to occupational illness claims.

In the first half of this century, insurers used a schedule rating system, which adjusted premiums to reflect the installation of safety equipment. Specific safe practices meant reduced premiums. Schedule rating brought about effective safety inspections by insurers. But, according to insurance industry representatives who testified at the hearings, the schedule system turned insurers into "policemen in the plants" and reduced cooperation from the companies in safety activity, so that it had to be abandoned.[46] Yet these same representatives were unable to present any hard data to show that experience rating had brought about an improvement.

Another area of uncertainty was the actual amount of the premium dollar spent by workmen's compensation insurers on job safety and health. According to Robert Heitzman, general manager of NCCI, general administrative expenses, which include safety expenditures, total 8.4 percent of the first one thousand dollars of premiums and 4.6 percent of premiums in excess of one thousand dollars. The NCCI recommends that 1.2 percent of this, or approximately $26.3 million, be spent for safety, but according to Mr. Heitzman, there is "no way to get the figures on what they [the insurers] actually spend because they are not required to report."[47]

Mr. Andrew Kalmykow, counsel to the American Insurance Association, testified at the hearings that casualty and property insurance companies spend $185 million annually for safety work that benefits employees.[48] When asked to give a breakdown of this figure, however, he revealed that only $32 million dealt directly with workmen's compensation insurance. The rest of the total comprised $75 million for fire prevention (including home safety), $31 million for auto safety, and $24 million each for boiler and machine safety, and for general liability (including elevator inspections).[49]

Standard-setting

Under any new job safety and health legislation, federal agencies would not have the manpower or the resources to fashion new standards for the myriad of processes and substances in industrial use. At least initially, therefore, the government would have to rely upon private standard-setting organizations. The Congressional hearings revealed that two major groups fashion safety and health standards: the American Conference of Governmental Industrial Hygienists (ACGIH) and the American National Standards Institute (ANSI).

Founded in 1938, ACGIH enables professional industrial hygienists working for the federal and state governments to accomplish what is functionally necessary for the implementation of their research. Frustrated within the federal bureaucracy, occupational health specialists use ACGIH as an outlet to devise standards the government has failed to set. Without ACGIH, there would be few guides to safe exposures to chemicals in the workplace environment.

ACGIH develops Threshold Limit Values, or TLV's, which are limitations on exposures to chemicals and physical agents beyond which workers should not be exposed over an eight-hour period. According to ACGIH, "Time-weighted averages permit excursions above the limit, provided they are compensated by equivalent excursions below the limit during the workday."[50] But for some substances ACGIH sets ceiling values that strictly limit exposure to that level and no higher at any time during the course of the day. The ACGIH has published a list of over four hundred TLV's. It takes about two years to develop a TLV.

The membership of the ACGIH closely follows the personnel roster of the Bureau of Occupational Safety and Health (BOSH). This organizational relationship fosters an extensive exchange of information, as well as a referral system for problems that may arise in either group. But its effect on BOSH is to divert any pent-up pressure that might be used on the HEW hierarchy to bring about a greater recognition of the importance of occupational health.

The American National Standards Institute (ANSI) used to be known as the United States of America Standards Institute (USASI). According to Ralph Nader, "The name was changed under pressure by the Federal Trade Commission who alleged that this was a deceptive practice in that it confused a private organization with a governmental agency."[51]

The organization has traditionally oriented itself toward the protection of industry from legal liability by creating standards that courts would adopt as governing corporate conduct.[52] More recently it has proclaimed its responsibility for the protection of the public, but without significantly changing its procedures to allow for meaningful representation of consumers and workers.

Reaching its fiftieth anniversary in 1968, ANSI provides a mechanism for the development of standards through the consensus method. Industrial safety standards constitute a small proportion of the more than three thousand ANSI standards, which range from specification for portland cement to size designations for index and record-keeping cards. There are about fifty-one of these job safety standards in common use, many of which have been incorporated into the Walsh-Healey Act.[53]

ANSI's membership falls into two categories: company members, composed of individual corporations; and member bodies, composed of trade associations, government agencies, and private groups such as the National Safety Council. Several labor unions have become member bodies within the past two years.

The institute is financed by membership fees and the sale of publications containing standards. About 80 percent of the membership fees come from company members who produce goods for which ANSI develops standards.[54] At one time ANSI placed a maximum on contributions from any one company. In 1970 the ceiling was lifted, enabling industry members to make contributions over and above their membership assessment. While this relieved some of the financial pressures on the organization, it also made ANSI vulnerable to what the maximum ceiling was designed to prevent—the domination of standards approval by major corporations.[55]

ANSI sets standards through standards committees drawn from its Member-Body Council. Prior to 1969, the council both developed and approved standards requested and paid for by industry. In 1969, the institute set up a separate body called the Board of Standards Review to create an aura of impartiality around the approval process. Of the fifteen members of this board, nine are from private industry, two from the federal government, one from municipal government, two from universities, and one from a consumer organization.[56] ANSI's president appoints the board, subject to the approval of the institute's Board of Directors, which also has authority to reverse any decision by the Board of Standards Review. Fewer than one-third of the directors were not directly tied to industry.[57] None were from labor organizations.

Individual companies do not directly participate in the standard-setting process, but if a member of the Member-Body Council happens to be employed by a particular firm, he will often check back with his company to see how he should vote.[58] Company members may suggest standards and initiate review proceedings or revisions of current standards. They are supposed to "represent the interests of United States industry and commerce in the activities of the institute and . . . promote the welfare of [ANSI]."[59]

Article 3 of ANSI's constitution defines consensus as "achieved according to the judgment of a duly appointed authority. Consensus implies much more than the concept of a simple majority but not necessarily unanimity."[60] The Board of Standards Review decides, by a two-thirds vote, whether a standards committee has reached a consensus. This decision is based not just upon the outcome of the voting in the standards committee, but rather upon the weight assigned to the votes of committee members. Therefore, while a majority consensus might exist, the negative vote of an important interested party might be given sufficient weight by the Board of Standards Review to amount to a veto of the proposed standard. All these determinations are made behind closed doors.

ANSI purports to balance its standards committees with a membership evenly divided among representatives of industry, government, and other interested parties. Committees usually have from twenty to seventy members. Since government, union, and consumer representatives are numerically in a minority, any "balancing" effect is pure illusion.

Thus, company members and their employees supply the technical expertise for the development of standards, which are then reviewed by representatives of trade associations (made up of these same companies), and finally passed upon by the Board of Standards Review, composed primarily of members from private industry.

The late Morris Kaplan of Consumers Union has described this incestuous process as follows:

> The industry representatives dominate the meeting. The consensus principle means in practice that the industry people have veto power over any action taken by the committee. . . . [The] object of the exercise is to get industry agreement, often arranged before the meeting anyway, and to push the standard through the paperwork of [ANSI] procedures so that it may be issued as a USA standard. Our proposals, or negative votes, are given "due deliberation," which is the phrase used in examining negative votes, but are ultimately vetoed or overridden as without merit. After a while it seems fruitless to spend the time and money to go to such meetings.[61]

A 1968 study by the Labor Department's Labor Standards Bureau found that nearly 60 percent of ANSI's standards were five or more years old, and more than half of these were over ten years old.[62] Prior to 1968, the average number of standards annually updated and revised was seventeen. This disclosure, combined with the Congressional hearings, spurred ANSI into action, and "during the first six months of 1969 the institute . . . approved 600 new and revised standards. This is more standards than were approved in 1968 and 348 more than in 1967, bringing the total number of standards up to 3,611."[63]

ANSI's laggard performance had aroused Congressional concern, since all the job safety and health legislation under consideration envisioned some role for private consensus standards. In addition, one particular ANSI standard came under heavy attack since it served as the measure for injury frequency statistics and was fingered as the cause of a massive misconception of the true magnitude of the industrial injury problem in the United States.

Injury-reporting statistics

The National Safety Council reported that in 1970 industrial accidents caused 14,200 deaths and 2.2 million disabling work injuries; 245 million man-days lost because of work accidents; $1.8 billion in wage losses; $900 million in medical costs; and a total cost to the economy in the amount of over $9 billion.[64] According to the U.S. Department of Labor's Bureau of Labor Statistics (BLS), in 1958 manufacturing entailed an accident rate of 11.4 disabling injuries per million man-hours worked. By 1967 the rate had reached 14, up from 13.6 in 1966.[65] Each of twenty-one major manufacturing groups surveyed by the bureau had a higher injury rate in 1967 than in 1965.

The measure that the BLS and the National Safety Council used to calculate disabling injuries was the so-called Z16.1 standard devised by ANSI. According to Z16.1, an injury is disabling, and hence reportable, if the "employee is unable, because of injury, to perform effectively throughout a full shift the essential functions of a regularly established job which is open and available to him."[66] This definition excludes an injury to an employee who can return to any "regularly established job" on the shift after an accident occurs. Thus, by reassigning injured workers to "soft" jobs, large firms easily avoid reporting injuries.

This is not a new ploy. Referring to a report made at the 1930 convention of the International Association of Industrial Accident Boards and Commissions, one writer has described "a fairly common practice for plants which are participating in accident-prevention drives or contests to give injured workers medical treatment and keep them in the plant on full wages rather than besmirch their 'no-accident' records by reporting them as lost-time accidents to the state."[67]

Exactly how this cover-up technique works came to light in testimony before the Senate Subcommittee on Labor. According to documentation supplied by the United Steelworkers of America, in a number of instances at a particular plant, injured workers were given different job assignments and ordered to return to work on crutches on the day after their accidents, while the company newsletter continued to boast about the company's record of consecutive hours without lost-time injuries.[68] Another union official, commenting on his company's receipt of an award for 2 million hours without a lost-time accident, said, "It's all walking wounded. That is what we call it."[69] According to Frank Burke, safety and health director of the United Steelworkers of America:

In the steel industry it is a practice that when an employee is injured they will administer first aid or hospitalization to the injured, they will have his or her supervisor punch their time card, then have another company representative go to the home of the injured employee and bring him or her to their place of employment so that this particular accident cannot be recorded as a lost-time accident.[70]

The Z16.1 standard excludes injuries incurred on a Friday if the employee makes it back to work on his next shift on Monday. Since the employee must be unable to work "throughout a full shift," the day of injury and the day on which the employee returns to full-time work are not counted as days lost as a result of disability.

Specific evidence of the type of minimizing inherent in the BLS and National Safety Council approach is observable in figures obtained from two large companies that calculated injuries on a different basis.[71] For December, 1968, Martin-Marietta listed eighty-nine "doctor cases," which they defined as involving more than "routine first aid." Over the same period they reported seven disabling injuries under the Z16.1 standard. For 1968, Bethlehem Steel reported 13.2 injuries per million man-hours of work according to its own "serious injury index," which included "all disabling work injuries (as defined in Z16.1) and all nondisabling work injuries which prevent the injured employee, for any part of a turn following the turn on which he was injured, from doing all or part of the job he was doing at the time of injury." For the same period the firm's National Safety Council rate was only 0.73 injuries per million man-hours. For 1967 Bethlehem Steel calculated a rate of 7.37 for its coal operations, but reported a frequency of 4.84 to the National Safety Council.

More than eleven years ago the chairman of the American National Standard Institute's Z16.1 Committee had this to say about the standard:

> As the safety movement developed . . . the emphasis on accident prevention tended to concentrate in particular groups of establishments, primarily the larger ones. . . . They took the lead in accident prevention and began to utilize all of the stimulants to effective action possible. This, inevitably, led to the competitive approach. . . . Unfortunately, for the statistician, these contests gave injury statistics in the individual establishment a position of importance which had not

been contemplated when Z16.1 was first developed. In effect, they lost their original purpose of measuring the need for accident prevention and became measures of accomplishment. . . . In all honesty, we have to recognize that most of the specific rules introduced into Z16.1 have the effect of reducing the range of reportable injuries. In the aggregate, the effect of these changes upon the range of reportable cases may be substantial. If we accept this premise, as I feel we must, all of our statistical indications of improvement in the volume of work injuries become questionable. Have we really succeeded in bringing injury occurrence in manufacturing to the lowest level in history, or do our figures largely reflect shifts in reporting rather than substantive improvement? Are we, in effect, kidding ourselves? If so, we are doing a disservice to the safety movement.[72]

The measurement of work injuries by the BLS suffers from other serious shortcomings. The BLS reckons its annual rate from data voluntarily submitted by some sixty-five thousand business establishments. A New York study that compared information voluntarily supplied to the BLS with information submitted by the same establishments, as required by law, to the state Workmen's Compensation Board revealed a considerable difference in work injury frequency rates. In manufacturing, the rate derived under the BLS approach was twelve accidents per million man-hours; as derived from the records of the New York Workmen's Compensation Board, it was sixteen.[73] For construction work, the discrepancy was twenty-seven to forty-three.

The BLS is well aware of inadequacies in its statistical approach. Indeed, an Alice-in-Wonderland passage in a BLS handbook attempts to turn night into day by confessing failure as an indication of success: "Whereas one might expect to breed a certain amount of doubt about a statistical survey revealing its lack of perfection, frankness about unavoidable defects more often has the opposite effect, and public confidence in the work is reinforced in the process."[74] Ralph Nader has quoted from a 1969 BLS report that took a somewhat less sanguine view: "Regional directors, industry safety people, and safety engineers were nearly unanimous in the opinion that Bureau of Labor Statistics figures were of little or no value."[75]

On June 30, 1970, Jerome B. Gordon, a safety and health consultant, submitted to the Department of Labor a report that concluded that injury statistics compiled by the BLS and the National Safety Council through the use of Z16.1 record a mere *one-tenth* of the actual injury toll![76] Gordon based his figures upon a serious-injury index that included nondisabling injuries that required medical attention beyond first-aid treatment.[77] When added to the evidence presented at the hearings, the Gordon study offered conclusive proof of the gross and unconscionable underreporting used by industry and its apologists for decades to cover up the epidemic proportions of the work safety problem in America.

Worker involvement

Most Congressional hearings take place in Washington. As the job safety and health hearings progressed, Senator Harrison Williams of New Jersey, chairman of the Senate Subcommittee on Labor, was persuaded to take to the road with his subcommittee in order to hear testimony from rank-and-file union members about local working conditions. In Jersey City, Pittsburgh, and Greenville, South Carolina, Senator Williams and his colleagues heard individual workers tell of the lack of information they had about the availability and results of state and federal inspections; cleanups that occurred because inspectors gave advance notice of inspections; health hazards workers believed they were exposed to; and the methods companies used to underreport injuries and illnesses. Fred Mann, president of Local 502 of the United Auto Workers, summed up the testimony of his fellow workers:

> Senators, after having been involved in the labor movement for seventeen years, and having seen throughout those years what I have, and having heard today what I have heard, it becomes very apparent that there is no other alternative but to say to employers through legislation, "This and this is what you must do."[78]

Rarely had unions brought members to testify at Congressional hearings. Indeed, it was only at the very end of the occupational safety and health hearings that the UAW, the Steelworkers, and the OCAW used this tactic effectively. (Miners afflicted by black lung, organized by the Black Lung Association and brought to Washington by Representative Ken Hechler of West Virginia, had haltingly wheezed their plight at hearings during the fight for passage of the Coal Mine Health and Safety Act and had made a tremendous impact on those Senators present.)[79] The injection of a flesh-and-blood dimension did much to prepare the way for passage of provisions that would directly involve workers in efforts to protect their own health and safety.

The Congressional hearings showed an undeniable need for federal job health and safety legislation. Industry and its supporters, however, fell back to a position of urging the passage of a weak bill. As 1969 drew to a close, the Javits-Ayres bill had obtained endorsement from a number of organizations that had bitterly opposed the Johnson administration's bill. Interest groups were maneuvering to preserve their stake in the health and safety "market" by making sure that their role would not be abolished or reduced by any new legislation. The trade associations, representing business, found they could give general support to the Nixon administration's bill. The Manufacturing Chemist's Association deemed the bill "acceptable."[80] The U.S. Chamber of Commerce, in 1968 the most intransigent and vitriolic opponent of federal legislation, opined that the Javits-Ayres bill "represents a responsible attempt to better our industrial safety and health record through an essentially sound program founded on federal, state, and business cooperation."[81] This sounded very much like a victory statement. If so, it was a bit premature.

The Battle for OSHA

The legislative history of the Occupational Safety and Health Act of 1970 (OSHA) provides more than just an arresting glimpse at the process by which laws are drafted and enacted. It also removes much of the gloss from the Nixon administration's blue-collar strategy. For while the President and Vice-President were making rhetorical pronouncements that purported to support working class aspirations, the Nixon administration was struggling to perpetrate a bill that would furnish no more than cosmetic protection for employees in an area that concerned them in the most vital ways imaginable.

Although the idea for a comprehensive federal bill originated during the Johnson administration, it took the Republicans to recognize and exploit the political potential of the occupational safety and health issue. But for the efforts of a few individual Congressmen and Senators, and the belated lobbying of a finally aroused labor movement, the Nixon administration might have succeeded in its attempts to water down the new law.

The original Johnson administration bill placed authority for standard-setting, inspection, and enforcement in the Department of Labor. This concentration of authority aroused all kinds of apprehension among business leaders, whose conception of the Department was that it catered to organized labor's every whim. This notion was fortified by the strongly worded testimony delivered by Labor Secretary W. Willard Wirtz at the 1968 Senate hearings:

> Mr. Chairman, S.2864 presents really one central issue, but that issue has as much to do with democracy as it has with health and safety. That issue . . . isn't states' rights, although we all know that S.2864 is going to be opposed by those who confuse that principle with their own interest.
>
> It will be a fair question of those who would oppose this bill whether it is their true belief that a comparable degree of responsibility may be expected from the states, and whether they will support the exercise of such responsibility—or hurry from opposing it here to some state capital to oppose it there. . . .
>
> The issue in S.2864 isn't cost. Although the issue is not the cost of protecting employees' safety and health, it will be opposed on that basis. . . .
>
> It is going to be a very fair question of those who oppose S.2864 on the basis of cost and expense to ask them just exactly what they regard as the price of a human life, or a limb, or an eye, and whether they would consider the price the same for a member of any family in America as they would for a member of their own family.
>
> So I say . . . that the clear central issue in S.2864 is simply whether the Congress is going to act to stop a carnage which continues for one reason, and one reason only, and that is because the people in this country don't realize what is involved. They can't see the blood on the food that they eat, on the things that they buy, and on the services they get. The question is democracy as well as health and safety.[1]

By 1970, the tone of the debate had changed. The new administration no longer considered that private industry and the states had failed but rather took the position that they needed a gentle shove from the federal government to get moving again. The picture of human drama that Secretary Wirtz had taken out from behind the statistics was replaced by a balance-sheet analysis of the costs of job safety and health. Secretary of Labor George Shultz's bland testimony to the same Senate subcommittee in 1970 reflected the change:

> We have observed the results of both private initiative and states' efforts, but we are aware of their uneven and imbalanced application. Our analysis of federal law shows that while federal responsibility can produce positive results, the piecemeal scope of federal authority limits its effectiveness.

The solution to these problems, we are convinced, lies in legislation which would assert a federal responsibility on a comprehensive scale for stimulating positive action and for placing a foundation under private and state efforts. Upon this foundation the federal government, the states, and private employers can build better programs to assure safe and healthful environments in which to work. The promulgation of basic federal safety and health standards will provide a useful starting point for state efforts.

In short, what is clearly needed is a broadly based federal law which will serve as a support for the work of the private sector and of the states, and which will involve them in the national concern for providing sound and healthful working conditions.[2]

The "broadly based law" to which Secretary Shultz was referring, the Javits-Ayres bill, gave the Labor Department authority over neither standard-setting nor enforcement. Fundamental to the Nixon administration's "game plan" for job safety and health was to vest this authority in an independent board appointed by the President. The unions vehemently opposed this. As a result, they directed their efforts against the board concept.

The unions were convinced that the independent board was a Trojan horse, an attempt to fragment any safety and health program and render it totally ineffective. A House Labor and Education Committee report succinctly stated the arguments against a board, and reflected labor's thinking on the subject:

The committee realizes that boards and commissions have been used in the past as a common technique to avoid making decisions, even where most of the information with which they deal has been readily available for direct Congressional action or administrative regulation.

A board whose members are appointed to serve for fixed terms could not be held accountable to anyone for reasonable and consistent establishment of standards. Indeed, it would be far better to place the authority in the one appointee whose primary obligation is to protect the legitimate interests of the workers and to enforce public policy in these areas as given to him by Congress and the President.[3]

One union lobbyist put it in more practical terms: "They [the unions] can get to the Labor Department, but not to a commission, and all protections are no good without standards." Union opposition to the board persisted throughout the legislative struggle, despite the antilabor performance of the Labor Department. Indeed, in October of 1970, the newly appointed Secretary of Labor James D. Hodgson, a former vice-president of Lockheed Aircraft, stated he would prefer no bill at all to the legislation being demanded by the AFL-CIO.[4]

In early 1970, the Democratic members of the House Select Subcommittee on Labor, led by Subcommittee Chairman Dominick V. Daniels of New Jersey, held a series of six informal meetings to consider the various bills that had been introduced prior to the 1969 House Hearings on Occupational Safety and Health and to hammer out a new bill that would be acceptable to the Democrats on the subcommittee. From these sessions and a number of staff meetings evolved a series of drafts, prepared by Daniel Krivit, counsel to the subcommittee, Gary B. Sellers, of Representative Phillip Burton's office, and George Skinner, a House legislative counsel.[5]

On March 13, 1970, Representative Daniels placed before the full subcommittee a revised draft that was to become known as the Daniels bill. A stronger version of the bill originally prepared by the Johnson administration in 1968, it countered what the Democrats saw as weaknesses in the Javits-Ayres bill by reducing reliance upon private standard-setting groups and placing authority for promulgating and enforcing standards in the Labor Department, rather than in an independent board. It also contained a significant innovation: guarantees that would involve workers themselves in efforts to safeguard the workplace.

During discussions by the subcommittee, the Republican members objected to much of the Daniels ·bill and were particularly critical of the delegation of both standard-setting and enforcement authority to the Department of Labor. In a gambit to prevent the bill from being reported out of the subcommittee, the Republicans stopped coming to meetings. There were nine Democratic members of the subcommittee, and it took the presence of nine members to create a quorum. The Republican boycott was based on the expectation that the Democrats would be unable to secure the attendance of all nine Democratic members on the day of the vote. In a counter-gambit, the Democrats called upon Representative Carl D. Perkins, Chairman of the full House Committee on Education and Labor and, by virtue of his position, a member of all his committee's subcommittees. On March 25, although one Democratic member of the subcommittee did not attend, Perkins's presence made a quorum and enabled the Democrats to report the Daniels bill out favorably, 9–0, at a meeting that no Republicans attended.

Before the full committee could vote on the bill, a more serious problem had to be confronted. Any legislation that was totally unacceptable to the Republicans on the committee would provoke a minority report and vigorous opposition—perhaps even defeat—on the House floor. The Republican members of the committee were working closely with the Labor Department, which was acting for the Nixon administration, and no issue of any import could be resolved without participation by the Labor Department. Therefore, Representative Daniels and then-Undersecretary of Labor James D. Hodgson arranged a special meeting in Republican Congressman William Steiger's office. Hodgson attended, accompanied by Solicitor of Labor Laurence Silberman. Steiger represented the Republicans on the House committee, and Representative William D. Hathaway appeared for the Democrats. Daniel Krivit, subcommittee counsel, represented Subcommittee Chairman Daniels, and Gary B. Sellers was present on behalf of California Congressman Burton. Over a three-day period in early June, 1970, the group discussed all of the approximately fifteen disputed areas, the main debate resulting in the acceptance of a board to set standards in return for substantive strengthening of worker protection in six to eight other areas. The group produced a compromise draft bill on which each side was to seek agreement from the other committee members, the administration, business, and labor. At the time of this meeting, the unions were still adamantly opposed to a board that would set standards because this would weaken authority and make all the other provisions just meaningless fringe benefits. They insisted that the Secretary of Labor had to have authority over both standard-setting and enforcement.

All those who took part in the meetings were aware of the unions' position on this issue, yet there was hope that labor could be induced to yield. But before the Democratic participants could approach organized labor and attempt to convince them of the merits of the compromise, union leaders found out about the meetings. Upset at their exclusion from these bargaining sessions, they concluded that the Democrats had committed themselves to a flabby compromise and accused them of a "sellout." According to Jack Sheehan of the Steelworkers Union, "Once the Democrats had accepted this compromise it made any battle in the Senate much harder." When the unions refused to go along with the compromise bill, a floor fight on the issue of the independent board became inevitable. But the unions' decision not to accept a compromise and to insist on the strongest possible bill did buy more time and also made the administration aware of how far it would have to yield to secure passage of a safety and health bill. On June 13, the full House Committee on Education and Labor adopted the Daniels bill over the strenuous objection of Republican committee members, who vowed to carry the fight to the House floor.

In September of 1970, Senator Peter Dominick of Colorado and Representative William Steiger of Wisconsin introduced a revised administration bill that contained more provisions for worker protection and less reliance on private standard-setting groups than previous Republican bills. But it divided authority even further. Standards were to be set by an independent National Occupational Safety and Health Board and would be enforced by an independent Occupational Safety and Health Appeals Commission. The Labor Department also shared enforcement responsibility, with authority to initiate complaints and make inspections.

The Dominick-Steiger bill marked a subtle shift in administration strategy to cope with the increasing pressure being brought to bear by organized labor. Its predecessor, the Javits-Ayres bill, had featured vague language and the complete absence of mandatory protections for workers. Administrative discretion had been the dominant motif, as espoused by the chief administration lobbyist, Solicitor of Labor Laurence Silberman, who seemed to be signaling to business that he could draft and administer legislation that would be even more lenient than existing laws.

Faithful to this policy, the Javits-Ayres bill had proposed

to give the Labor Department and the independent board complete discretion on whether or not to act in protecting the workers' safety and health. The board would be allowed to decide whether to set standards at all. The Labor Department had complete discretion to ignore violations found during inspections; or if violations were found, to petition the board for a hearing to determine whether the violation was punishable. Of course, the Labor Department had no duty to make any inspections at all:

> If upon inspection or investigation, the Secretary within his discretion determines that there is reasonable cause to believe that an employer has violated standards . . . he may petition the board for a hearing to determine if a violation has occurred.[6]

The only penalty was a ten-thousand-dollar fine for a willful violation, which the board could decide not to assess. Every violation in a plant required a full administrative hearing that could have taken years. The bill was worse than no bill at all, for it would have foisted on the workers a cumbersome bureaucratic machine that could choose to do absolutely nothing.

The second administration proposal, the Dominick-Steiger bill, introduced thirteen months later, was described as a "bipartisan" solution to the safety and health problem only because the House version was co-sponsored by Democratic Congressman Robert Sikes of Florida. It amounted to a carefully fashioned attempt to win the votes of Democratic Congressmen from sections of the country where organized labor was weak.

Except for its tripartite division of authority to set and enforce standards, the Dominick-Steiger bill represented a distinct improvement over the Javits-Ayres bill. Indeed, in many ways it approximated the compromise draft rejected by labor earlier in the year. But from the perspective of optimum safeguards for workers, the Daniels bill remained far superior. The Dominick-Steiger bill seemed to ignore all the testimony at the hearings about deadly gases, dusts, and chemicals. It contained few guidelines for the measurement of the adequacy of health regulations, no mandatory provisions for monitoring or measuring exposures to toxic materials, no provisions for medical examinations or employee access to personal files. HEW's role was vague and feeble. Its occupational health program received no new status, and the independent board, rather than HEW, was given authority over employee requests regarding toxic substances. Aside from inspections, the Secretary of HEW had no duty to gather exposure data or conduct industry-wide studies to ascertain the scope of specific health problems. The Daniels bill required the Secretary of Labor to promulgate national consensus standards within two years of the effective date of the law; the Dominick-Steiger bill stretched this period to three years. All in all, language in the Dominick-Steiger bill was always couched in somewhat weaker terms than that in the Daniels bill.

In late summer and early fall of 1970, the Senate Labor Subcommittee began work revising the strong bill introduced in 1969 by the subcommittee's chairman, Senator Harrison Williams of New Jersey. The subcommittee did not split along party lines, since several key Republican members—notably Senators Javits of New York and Richard S. Schweiker of Pennsylvania—came from states with large working-class populations and hence were more sensitive to pressures from organized labor. Though Senator Dominick pressed for the adoption of the Dominick-Steiger bill, the subcommittee turned him down and inserted into the Williams bill several provisions suggested by Senators Javits and Schweiker. But Senator Javits's attempt to incorporate an independent board failed, so that authority both to set and enforce standards remained vested in the Secretary of Labor. On September 9, the subcommittee reported out the new version of the Williams bill, which was adopted by the full Senate Committee on Labor and Public Welfare on September 25.

The Williams bill did not reach the floor of the Senate until November 16, shortly after the 1970 election recess. Senator Dominick again offered his bill as a substitute, and lost by only two votes. Senator Javits then introduced an amendment to create a three-member Occupational Safety and Health Review Panel to exercise enforcement authority. With Southern Democrats joining a solid phalanx of Republicans, and five pro-labor Democrats—Senators Birch Bayh, Alan Bible, Eugene McCarthy, Clairborne Pell, and Joseph Tydings—missing, it passed by a margin of forty-three to thirty-eight. (According to

George Taylor of the AFL-CIO, three liberal Republicans would have switched votes to oppose the review panel if the Democrats could have mustered the five votes necessary for a tie.) With this basic compromise inserted, the bill was enacted eighty-three to three, Senators James O. Eastland, Sam J. Ervin, and Strom Thurmond remaining adamantly in opposition.

A week later, the Daniels bill reached the floor of the House. Congressmen Steiger and Sikes introduced their bill as a substitute, and the battle lines were drawn. Some compromise seemed necessary if the Republican strategy was to be defeated, but the Democrats found themselves boxed into a corner. Representative Daniels was closely coordinating his moves with the wishes of the AFL-CIO and other unions. While Daniels could privately indicate to his fellow Congressmen that he was willing to soften his bill in line with the Senate version (including acceptance of a board to review citations), the decision was made that Daniels would not announce this in public. The AFL-CIO was urged to make a public statement of its willingness to compromise, but refused. All Daniels could do was hint on the floor of the House that he would be willing to amend his bill if his colleagues would first vote down the Steiger substitute:

> We are not too proud to improve our bill and, in order to make every concession to alleviate fears that have been expressed, we will propose modifications to the committee bill that, while not interfering with its effectiveness, will reduce areas of concern that have been expressed.[7]

In the last few hours the union lobbyists finally conceded the need to take a candid position, and Daniels introduced a series of specific amendments he would accept. These did not include a review board amendment, the main issue at this point. His effort turned out to be too little and too late. The administration bill, since it was offered as a substitute, came to a vote first, and passed, 220–172. Southern Democrats and liberal Republicans who might have voted for a compromise found themselves forced to choose between the Daniels bill and the Steiger substitute. Between these two extremes, they opted for the latter, which was less controversial. At this point, each side had scored a partial victory. Labor had won in the Senate, and the administration had won in the House. For Representative Daniels, however, this marked a bitter disappointment, for it meant that despite his seminal efforts, the bill would not bear his name.

When the Senate and House pass conflicting versions of a bill, a final version is hammered out by a joint conference committee, composed of members of both parties from both chambers, and is then sent back to the House and Senate for final passage (or rejection). After nearly three years of testimony, lobbying, and meetings, a conference committee would now decide the fate of the job safety and health bill in the course of several weeks. These conference sessions became crucial.

During the conference itself, which took place in December of 1970, there were several levels of operation and a series of outside influences affecting the participants.

The Democratic House members on the committee were in the awkward position of having to support a bill they had strenuously opposed. They could not completely abandon the Steiger substitute, as this would mean certain defeat for the conference bill in the House. Representative Perkins, the chairman of the House Labor Committee, and Representative Daniels led the Democratic House contingent on the conference committee. They had to evaluate with great care each area of disagreement between the two versions, and decide when they should accept provisions from the stronger Senate bill, and when they had to vote with their Republican colleagues for a weaker provision in order to secure passage in the House. Perkins, who was selected chairman of the conference committee, was in a particularly difficult position. He was at times subjected to strenuous pressure from Congressmen Burton and Hathaway to hold firm for the Senate bill, since the Democrats had the majority in the conference committee if they stood together. But Perkins handled himself well throughout the proceedings. For example, on the issue of formal-vs.-informal rulemaking, the Democrats accepted the Senate provision for informal procedures, but only after extensive debate and soul-searching between Perkins and the other Democrats. The latter maintained that the only way to administer the law was by informal rulemaking, and that it would be too long and arduous a process to require the Labor Department and the unions to go through formal procedures.

The Republican minority on the committee deserved an Oscar. They had put up such resistance all along that most Democratic staff personnel were certain they would never approve a conference report. Without some Republican support, most observers believed that the conference bill could never receive a majority in the House and that even if it did pass, the President would veto it. The Republicans fostered this impression and complained as vigorously as they could each time a majority on the committee accepted one of the Senate bill's provisions. What they succeeded in concealing was a decision on the part of the administration not to torpedo the job safety and health bill. This amounted to a tacit Presidential recognition of the volatile political potential of the occupational environment issue. The administration did not want to be put in the position of being responsible for blocking the bill. Thus, to the Republicans, the substance of the bill as enacted became relatively unimportant; what was essential was for the Republicans to be able to say, as part of their blue-collar strategy, that the Nixon administration was doing something for the workers. Most of the Democrats wanted a strong bill even more than they wanted to avoid a Congressional defeat. But the Republicans, by referring constantly to the possibility of their refusal to sign the conference report if the bill were too strong, were able to win concessions in the conference.

Within this atmosphere of countervailing pressures and general uncertainty, the potential role of individual members of Congress and lobbyists was far greater than usual. The key to many of the conference decisions can be found in their maneuvering. The business community had several men outside the door throughout the conference. They were: William Wickert of Bethlehem Steel, George Burnham of U. S. Steel, Anthony Obidahl of the U. S. Chamber of Commerce, Tom Mitchell of Georgia-Pacific Forest Products, Scott Shotwell of the Association of General Contractors, Walter Jaenicke of the National Forest Products Association, and Harry Rosenfield of the National Safety Council. For labor, Jacob Clayman represented the Industrial Union Department of the AFL-CIO; Jack Sheehan, the United Steelworkers; Anthony Mazzocchi, the Oil, Chemical and Atomic Workers; and George Taylor and Howard

McGuigan the AFL-CIO. Labor Department lawyers presented the administration views through the new Solicitor, Peter Nash, and his associate, Henry Rose. HEW sent Pat Foley, the administrative assistant to the director of the Bureau of Occupational Safety and Health, but no top personnel.

All the pressure from the unions bore down on the issue of who should set standards. Before the conference had begun, the union representatives, led by Jack Sheehan, Anthony Mazzocchi, Arnold Mayer of the Meat Cutters Union, and George Taylor of the AFL-CIO, had decided that what they wanted above all else was that the Department of Labor, and not a board, set the standards. In order to reach this goal, the unions were willing to make certain concessions, which they conveyed to Senator Williams and Representative Perkins. For example, the unions agreed that an inspector, upon discovering an imminent hazard, would have to obtain a court restraining order and could not temporarily shut down a plant on the spot, as authorized by the Williams bill. This compromise was incorporated into the final bill.

What was most apparent throughout these sessions was the role of the Department of Labor officials as spokesmen for business. The unions recognized this and accepted it, hoping for better luck in future administrations. The Department's representative, Solicitor Peter Nash, supplied amendments on behalf of the administration to the Republican conferees, all of which were directed toward weakening the bill in line with proposals by business interests.

HEW's role during the conference remained one of nonperformance, even though the Labor Department was offering amendments to weaken HEW's role, delete its authority, transfer its functions, and otherwise emasculate the strong health provisions contained in the Senate bill. For example, the Senate bill stated that an employer could be required to monitor potentially toxic substances for which no exposure limit had been set, and to allow employee access to the results of this monitoring. A Republican amendment was accepted that permitted the Secretary of HEW to monitor potentially toxic substances and physical agents in order to obtain information for standards, but did not explicitly permit employees to look at records of their own exposures to substances or physical agents that were potentially toxic. Thus, the only exposure records to which employees now have a specific right of access under the act are those involving substances or physical agents that are known to be actually toxic.

The issue of the independent board gradually resolved itself as a result of union concessions. The Senate proposal of a review panel (now called a review commission) was adopted, with standard-setting authority itself given to the Secretary of Labor. This left as key issues the method of reporting injuries and diseases, and the problem of inspections. Injury statistics had been based on the Z16.1 measure developed by the American National Standards Institute (ANSI). This measure had led to a variety of forms of underreporting, as described in Chapter 7. The bills before the conference required all injuries to be reported—a legislative mandate so broad that both industry and the Labor Department had said that they would be neither able nor willing to comply. But Democratic staff members realized that this impasse could be used to the advantage of the workers and drafted their own amendment. Congressman Burton, pointing out that the existing provisions were much too broad, proposed language that required the reporting of all work-related injuries "other than minor injuries that do not involve medical treatment, loss of consciousness, restriction of work or motion, or transfer to another job."[8] This put some limitation on the injuries to be reported, but also cured the major defects in the ANSI standard.

Not every proposal turned out so successfully. The Senate bill had a self-inspection provision that obliged the employer to certify the results of his inspection to the Secretary of Labor. No one, however, had introduced an amendment calling for a minimum number of mandatory government inspections. This was a sensitive issue since, under existing law, inspections depended upon the resources available and therefore affected only a small percentage of establishments. Representative Burton asked his assistant to inquire whether Solicitor of Labor Nash would accept a minimum inspection provision. But the Labor Department would neither introduce nor support such an amendment. Indeed, even the Senate provision for certification of self-inspection results was dropped from the final version of the bill.

The conference committee finally succeeded in reporting out

a bill acceptable to both sides, and on December 17, 1970, it came up for a vote in both the Senate and House. The Occupational Safety and Health Act (also known as the Williams-Steiger Act) passed by wide margins; and on December 29, 1970, President Nixon signed it into law. The Republicans had their political drawing card, and the unions had a significant piece of legislation that could be used to safeguard the industrial environment.

The passage of the Occupational Safety and Health Act (OSHA) substantially changed the responsibilities of organizations, agencies, and individuals dealing with job safety and health. Within a rather complicated regulatory framework, OSHA affects in varying degrees the states, the Departments of Labor and of Health, Education and Welfare, the federal safety program, labor unions, and individual workers.

Impact on the states

Though the record that the states have compiled on matters of occupational safety and health is on the whole a sorry one, the new federal statute has by no means eliminated state involvement. Quite the opposite. OSHA provides, first, that state health and safety codes covering areas unaffected by federal standards will remain in effect. For example, since no federal standards dealing with boiler or elevator safety have ever been issued, state laws and regulations on boilers and elevators remain in force.

Second, the Secretary of Labor may allow the states to continue to enforce some of their own safety and health standards for a period of up to two years from OSHA's enactment. This, of course, would not affect the promulgation and enforcement of federal standards under the act.

Third, OSHA authorizes the states to regain responsibility for safety and health by putting together state plans for the regulation of the occupational environment. The act provides for a grant-in-aid program that will fund the development, and later, the administration, of state plans.

In order to obtain federal money and federal approval for state plans, and thus avoid federal preemption of job safety and health regulation, the states must draw up and submit their plans to the Labor Department within two years of the signing of the act, and must meet several criteria spelled out in the act and in Labor Department regulations. Basically, a state plan covering one or more areas of safety and health will be approved if the Secretary of Labor determines that it "is or will be at least as effective" as the federal act. The areas where the state must provide the Labor Department with data on effectiveness include: development and enforcement of standards, right of entry and inspection, legal authority and qualified personnel, and funding. Submission of plans to the Department of Labor does not end the states' responsibilities under OSHA. The public may comment on the state proposals, which will be made available for inspection. Even after the Labor Department has approved a state plan, the Department has discretionary authority to enforce federal safety and health standards within the state for at least three years. As interpreted in the Department's regulations, this authority for continuing enforcement will be used when the state plan, as approved, will not be complete until a future date (when, for example, additional state legislation is needed to meet federal requirements).

Even after all criteria have been met and the Labor Department can no longer enforce federal standards in areas governed by the state plan, the state will continue to be subject to scrutiny by the federal government—to assure that the states do not backslide. If a state fails to comply substantially with its own plan, the Secretary of Labor must withdraw his approval after a hearing.

Occupational safety and health administration

The main impact of the new act directed itself at the Department of Labor. Under a new Assistant Secretary for Occupational Safety and Health, the Department is charged with assuring "so far as possible, every working man and woman in the nation safe and healthful working conditions."[9] The act spells out in detail how safety and health standards are to be set, inspections made, and citations for violations issued. In addition to that enforcement authority, the Department also must provide training and education programs for safety and health personnel, as well as for employees and employers, and must gather information on the extent of occupational injuries and diseases.

The new Assistant Secretary, under delegation of authority from the Secretary of Labor, has authority to set three types of standards. The first, designed for rapid promulgation within a two-year period, derive from the national consensus standards of the ANSI and from federally established standards under the Walsh-Healey Act, such as the Threshold Limit Values of the ACGIH. These consensus standards are to be promulgated without a hearing or public comment, since they presumably allowed for input from interested groups when they were originally developed.

A second type of standard is subject to a more formal proceeding, including a hearing. These are standards for toxic materials or harmful physical agents; the standards must be designed to assure "to the extent feasible, on the basis of the best available evidence, that no employee will suffer material impairment of health or functional capacity, even if such an employee has regular exposure to the hazard dealt with by such standard for the period of his working life."[10]

Third, because it may take a substantial amount of time to promulgate the second type of standard, the act provides for emergency temporary standards that furnish immediate protection, subject to a full hearing to develop any warnings or monitoring requirements.

In addition, OSHA provides that all of these new standards should contain labeling requirements and warnings for employees concerning hazards, symptoms, precautions, and emergency treatment. The Department may include provisions for monitoring employee exposure and for medical examinations. By listing these authorizations in the act itself, Congress has specifically ordered the Labor Department to build preventive techniques into the standard-setting process.

Inspection authority is a necessary concomitant to standards in order to check on compliance by employers. The Department of Labor has specific authority for safety and health inspections, including the right to talk privately with employers or employees. The compliance officers of the Department may also inspect safety and health records of the employer. The Department has the authority to issue citations listing violations found during inspections and to assess penalties against the employer. The citations include a time period for corrections. Either the employer or the employees can contest them before the Occupational Safety and Health Review Commission. To keep employees informed of alleged violations in the workplace, the employer is required to post citations at or near each place of violations.

Violations are classified as imminent danger, serious, and nonserious. An imminent-danger situation is one where "a danger exists which could reasonably be expected to cause death or serious physical harm immediately or before the imminence of such danger can be eliminated through the enforcement procedures otherwise provided by this Act."[11] In such a case, the Secretary of Labor may ask a federal district court for a temporary restraining order to shut down the dangerous operation. A serious violation exists if:

> . . . there is a substantial probability that death or serious physical harm could result from a condition which exists, or from one or more practices, means, methods, operations, or processes which have been adopted or are in use, in such place of employment unless the employer did not, and could not with the exercise of reasonable diligence, know of the presence of the violation.[12]

A nonserious violation is any violation that is determined not to fall within the terms of the definition of "serious violation." The distinction is significant in terms of whether penalties have to be assessed. Penalties are mandatory if a violation is serious, but discretionary in nonserious cases.

Enforcement is only one of the purposes of OSHA. Education plays a role, too, and the Department of Labor is authorized to develop short-term training for safety and health personnel. In conjunction with HEW, the Department must also educate and consult with employers and employees on the recognition, avoidance, and prevention of occupational hazards. In order to set priorities in enforcement, education, and prevention, the Department is required to set up an injury and illness reporting system.

An end to BOSH

While the Department of Labor received overall supervisory authority under OSHA, HEW still has a role to play. Thanks to one of Senator Javits's amendments, the act elevates the Bureau of Occupational Safety and Health (BOSH) to the level of a National Institute (NIOSH) within HEW, in an attempt to raise the visibility and importance of safety and health within the HEW bureaucracy. In addition, the law prescribes certain mandatory functions for the new institute. It must prepare an annual list of toxic substances and conduct industry-wide studies of the "effect of chronic or low-level exposure to industrial materials, processes, and stresses on the potential for illnesses, diseases, or loss of functional capacity in aging adults."[13]

This research is to be used as background information for the setting of health standards. BOSH had never directly participated in the development of standards, but when the Labor Department now decides to issue, or is petitioned to issue, a standard in the occupational health area, the new institute's research will provide the basis for the standard. Thus, OSHA attempts to mobilize the long-dormant occupational health mission of HEW.

After Mission Safety-70

The Occupational Safety and Health Act extends federal standards to the various federal agencies and departments. Enforcement provisions and sanctions are not applicable, however, although the agency and department heads, in the words of the act, "shall" provide safe and healthful workplaces, procure safety equipment, keep records, and make annual reports. In effect, MS-70 will become a permanent program, giving the Federal Safety Council received statutory authority and a clearer definition of its responsibilities.

Since the new federal standards apply to all agencies and departments, the government employee unions can play an important role that transcends mere consulting with the FSC. Members of these unions are now represented on the Federal Safety

Advisory Committee. They can make certain that the federal government complies with safety and health standards, provisions for monitoring, medical examinations, and recording of exposure levels under the new standards. This would go a long way toward securing for federal employees a safe and healthful workplace.

Employee rights under OSHA

The signal accomplishment of the Occupational Safety and Health Act is that it explicitly recognizes the role of the individual worker and the union locals in the safety and health area. In the standard-setting process, the Secretary of Labor must "promulgate the standard which assures the greatest protection of the safety or health of the affected employees."[14] Any worker or other interested person has the right to submit data and written comments during these procedures. Without waiting for government action, workers or their representatives can initiate a rule-making proceeding by a written petition to the Secretary of Labor. Labor representatives have the right to sit on advisory committees that consider the new standards. If an individual employer seeks a variance from any promulgated standards, he is required to inform his workers, post details, and make copies of his petition available. This information must also inform the worker of his right to oppose the variance at a hearing.

The information gap separating employers and employees is partially bridged by provisions for regulations requiring employers to inform workers of their protections and obligations under the act, including the content of applicable standards. Labor Department regulations provide that all citations for violations of standards must be prominently posted near the site of the violation. In any enforcement proceeding before the Occupational Safety and Health Review Commission, an employee or his representative has the opportunity to participate to show that the Secretary of Labor, in issuing the citation that the employer is challenging, gave the employer an unreasonably long time to correct a violation of a standard. Finally, as a person "adversely affected or aggrieved" by an adverse decision of the commission, an employee can appeal to the courts to modify or set aside the commission order. The act forbids employers from discharging or discriminating against employees who exercise any of these rights.

A crucial factor in the debates over occupational health is the lack of medical data on the extent and effect of harmful exposures. In addition, workers were, and are, often unable to obtain any information about the chemicals they are exposed to —preventing them from obtaining necessary medical treatment and limiting compensation payments if they cannot establish that their illness was work-connected. Section 8(c)(3) attempts to reach these problems by providing that employers record exposures and give their employees the opportunity to observe the monitoring process. Only if a substance is defined as toxic (that is, if the Secretary of HEW includes it in his annual list of toxic substances) and is covered by a standard is the employee entitled to information about his exposure record. If the Secretary has not listed a particular substance, an employer or an authorized representative of an employee (but not an individual worker) may request the Secretary of HEW to determine "whether any substance normally found in the place of employment has potentially toxic effects in such concentrations as used or found."[15] If HEW makes this finding, the Secretary of Labor may initiate a rulemaking procedure to set a safe exposure level for this substance.

The only exposures that must be monitored are those for which the Secretary of Labor has set exposure standards, and those whose standards contain specific provisions requiring monitoring. Therefore, if a substance has been determined to be toxic, or to be a harmful physical agent, and the standard includes a provision for monitoring, then the employer, according to regulations prescribed by the Secretary of Labor, shall make "appropriate provision for each employee or former employee to have access to such records as will indicate his own exposure to toxic materials or harmful physical agents."[16] The employer is also to notify an employee whenever his exposure exceeds that prescribed in the standard.

The Secretary of Labor is authorized, but not required, to inspect workplaces covered by the act. If the Walsh-Healey tradition holds, over 95 percent of the workplaces will never be visited. If an inspection does take place, OSHA gives the employee's representative the opportunity to accompany the inspector. And if an employee or his representative knows of an existing violation prior to or during an inspection, he may notify the inspector orally or in writing. If the inspector does not issue a citation on the basis of a written request, the employee or his representative may ask the Department of Labor for an informal review to find out the reason. Labor Department regulations have defined "employee representative" to include any person who the inspector determines will aid in the inspection process.[17]

Although the Department of Labor refused to ask Congress for a mandatory inspection system in the act, there is a provision authorizing the Secretary, at his discretion, to promulgate regulations requiring periodic self-inspections by employers. In the Senate bill, the employer had to certify the results of such inspections to the Secretary. This provision was dropped in conference. The Secretary of Labor can nonetheless require employers to include information on these inspections (but not specific violations) in their reports to the Department of Labor.

Since any federal inspection system is likely to be minimal, it is important that the unions and individual employees recognize the potential of Section 8(f)(1), which states, "Any employees [more than one is required] or representative of employees, who believe that a violation of a safety or health standard exists that threatens physical harm, or that an imminent danger exists, may request an inspection." The employees have to give detailed reasons and provide copies for the employer. This can be done with or without disclosing the name of the employees making the request. Upon receipt of the request, the Secretary of Labor must respond by determining whether reasonable grounds exist to believe there is a violation. If his finding is in the negative, he must say so in writing. This would appear to constitute a "final administrative order," which would make it appealable to the courts. If the Secretary does find reasonable grounds, a special inspection shall be made "as soon as practicable."

A violation that threatens physical harm would arise, for example, if a standard had been promulgated concerning nonremovable guards on machinery. If the employer did not install such guards, or installed guards that were removable, then a violation would exist. Its existence would threaten physical harm, e.g., the amputation of a hand. (The violation does not have to threaten *serious* physical harm, and the provision also applies to health violations.) The Secretary of Labor, on receiving an inspection request, would not have to find the existence of a violation, but merely reasonable grounds to believe a violation exists.

The second situation deals with imminent danger, a provision that was contested throughout the passage of the bill. While physical harm is a relatively simple concept, the definition of what constitutes an imminent danger is more elusive. In addition, a finding that an imminent danger exists will result in a much more serious penalty than a mere citation, for the Secretary must apply for a court order that would restrain the conditions or practices creating the imminent danger. This provides another weapon for employees, but it depends on an inspection by a Labor Department official and a petition by the Secretary on the basis of that inspection. If the Secretary "arbitrarily and capriciously" refuses to petition the court after receiving the results of an inspection that revealed the existence of an imminent danger, OSHA provides that the employees or their representative can bring a court action for mandamus against the Secretary to force him to act.

The statutory definition of imminent danger has already been quoted. One problem is how immediate the risk of death or serious harm must be. One section of the definition posits that before declaring the existence of an imminent danger, an inspector would already have to be on the premises, and the serious injury would have to be likely to occur reasonably soon. Under the second clause of the definition, imminent danger can also mean a hazard that cannot be eliminated before the full enforcement procedure can be followed. Thus if the employees can show that consistently high levels of a toxic substance could reasonably be expected to cause serious physical harm that was irreversible, although not manifest at once, the employees could obtain a special inspection.

The Assistant Secretary of Labor for Occupational Safety and Health, in a letter to Dr. Donald Whorton of the Health Research Group, has acknowledged that the imminent-danger provision can apply to health hazards. According to the Assistant Secretary, "Exact conditions and circumstances are difficult to describe for such conditions. Generally, such would be for acutely-acting and/or systemic toxins and some extremely high level of physical agents."[18]

OSHA gives workers and their unions an often complex but still workable tool for their own protection. The development of the act moved from the establishing of conventional regulatory authority to the fashioning of statutory guarantees that enable workers to participate in the enforcement mechanisms as well as the administrative process.

Workmen's compensation

One final section of the act is worthy of note. Thanks to the initiative of Senator Javits, OSHA created a National Commission on State Workmen's Compensation Laws, which was charged with conducting "an effective study and objective evaluation of state workmen's compensation laws in order to determine if such laws provide an adequate, prompt, and equitable system of compensation for injury or death arising out of or in the course of employment."[19] The commission was charged with completing its study by July 31, 1972.

The First Year of OSHA

When President Nixon signed the Occupational Safety and Health Act on December 29, 1970, he hailed it as "perhaps the most important piece of legislation to pass in this Congress."[1] OSHA had transformed the federal government's responsibilities for the protection of the workplace environment from a walk-on part to a leading role. However, as is so often the case, OSHA's true worth lies less in the script itself than in the way the actors play their roles. If the first performances are any guide, OSHA is in trouble.

Although OSHA has been in effect for but a short time, enough of a record has been compiled to permit an assessment of three vital aspects of this new phase in the history of job safety and health regulation: the performance of the Departments of Labor and HEW; OSHA's impact upon state regulation; and organized labor's response to the new law and its administration. Other developments worthy of note include the controversy over the costs of OSHA, its impact upon small businesses, and the final report of the National Commission on State Workmen's Compensation Laws.

Appointments: old wine in new bottles

OSHA provides a novel, complex, and sophisticated mechanism for dealing with a problem of massive dimensions. It cries out for creative, dynamic administration, and this would seem to require an infusion of new talent in key positions. And yet a perusal of the new job safety and health slots in the Labor Department and HEW reveals a roster of familiar faces—with promotions.

In mid-March, 1971, President Nixon announced the appointment of George C. Guenther as Assistant Secretary of Labor for Occupational Safety and Health. Formerly president of the John H. Guenther Hosiery Co. of Reading, Pennsylvania, Mr. Guenther had been serving as chief of the Labor Department's Bureau of Labor Standards. A handsome, photogenic man with a penchant for cigars, the new Assistant Secretary did not bring a firm hand to his new post. The office of the Associate Solicitor for Occupational Safety and Health does most of the work drawing up the regulations implementing OSHA, with close and constant scrutiny from newly appointed Undersecretary of Labor Laurence Silberman. Thus, the former Solicitor, who had been a key lobbyist for the weak administration bill, now enjoys supervisory authority over the administration of a law he tried to gut.

The appointment of Mr. Guenther evidenced an upgrading of the Bureau of Labor Standards within the Occupational Safety and Health Administration and closer coordination with the mainstream of the Labor Department. But it did not indicate a change in policy. During his less-than-thorough grilling on the Labor Department's plans at his *pro forma* Senate nomination hearings, Guenther emphasized fairness to business. He mentioned the unions only once, and made no reference to the workers. His subsequent appointments reflected this orientation. It was only after persistent criticism from the unions that he finally named a union man (Maywood Boggs of the AFL-CIO's Metal Trades Department) as one of his special assistants.

Simultaneous with the announcement of the Guenther appointment, the White House also made public the names of the three men who would serve on the Occupational Safety and Health Review Commission. Robert Moran, from the Labor Department's Workplace Standards Administration, had worked diligently in organizing task forces to do some of the preliminary work under OSHA. He was rewarded with a six-year term as commission chairman. James Van Namee, safety director of Westinghouse, and Alan Burch, of the Operating Engineers Union, completed the lineup.

These choices typify the functional labor-management-government breakdown that characterizes the appointment process. Senator Javits, in a debate over the appropriate criteria for appointing members of the commission, had argued that "training, education, and experience are good criteria rather than pinning it down to one from labor, one from management, and one from the public."[2] Section 12(a) of the act does specify that commission members be picked on the basis of "training, education, or experience," but these guidelines were still not specific enough to prevent the administration, in a Pavlovian exercise of "discretion," from resorting to the traditional troika of labor, management, and public.

The Senate Labor and Public Welfare Committee nomination-hearing transcript reveals that almost no questions were put to the review commission nominees. Several Senators stated they were satisfied on the basis of private chats in their offices with the nominees—outright sabotage of the main purpose of a public hearing. (If the Senators didn't want to ask any additional questions in public, they should at least have included in the hearing transcript a summary of their private discussions.)

The unions were not contacted about any of these appointments. As George Taylor of the AFL-CIO stated in February, 1971, in a radio interview, "Well, we're not privy to the inner workings of the Nixon administration. We have no input as far as our people are concerned."[3] Complete lack of consultation with what is supposed to be its constituency—the unions and their members—demonstrated once again that the Department of Labor under President Nixon had no intention of changing its past attitudes toward the working man.

OSHA also provides for a National Advisory Committee on Occupational Safety and Health (NACOSH). The men and women who would staff this advisory body were to be selected on the basis of their "experience and competence in the field of occupational safety and health."[4] Their advice is meant to guide the policy-makers and civil servants. OSHA specifies that NACOSH shall consist of twelve members, eight appointed by the Secretary of Labor and four by HEW, all representing management, labor, occupational safety and health professionals, and the public—with a public member as chairman. The critical balance was to be held by the public and professional members.

Secretary of Labor Hodgson announced the members of NACOSH in late July, 1971. Not surprisingly, the deck was stacked in favor of business. The lineup on paper is six industry or industry-oriented members, with two labor people, and four members nominally not affiliated with industry (although two had industry backgrounds). Howard Pyle, president of the National Safety Council, is the chairman, even though he comes from a recognized industry-oriented group, not a public organization as required by the law. Mr. Pyle's chairmanship further unites the administration of OSHA with the dubious safety and health policies of the National Safety Council. Representing industry, in addition to Mr. Pyle, are Frank Barnako, manager of safety and workmen's compensation for Bethlehem Steel and a member of the Board of Directors of the National Safety Council; Roscoe Batts, vice-president for industrial relations of International Harvester Co.; Donald Peyton, managing director of the American National Standards Institute, an industry-funded organization that develops standards for submission to the Labor Department for consideration by the committee on which Mr. Peyton sits; Sara P. Wagner, director of nurses for Standard Oil of New Jersey; and Roger Wingate, vice-president of the Liberty Mutual Insurance Co. of Boston.

The four members who have no industry ties are Barry Brown, director, Michigan Department of Labor (formerly an attorney and executive for Chrysler Corp. and the J. L. Hudson Co.); John Grimaldi of the Center for Safety at New York University; Richard Sutter of the Sutter Clinic in St. Louis, Missouri; and Lief Thorne-Thomsen, Public Health Administrator, Department of Health and Welfare, Alaska.

Bringing up the rear (and symbolically seated at the foot of the table at the first NACOSH meeting) are George Taylor of the AFL-CIO Standing Committee on Occupational Safety and Health, and John Sheehan, legislative director of the United Steelworkers of America, both of whom are essentially union lobbyists with much knowledge in the job safety and health field but lacking the professional credentials that would enable them to counteract the probusiness bias of the Labor Department. They enjoy no support from the other members of the committee and merely play the role of nettles in the side of the Labor Department. In sum, NACOSH is a group of predominantly like-minded men and a woman more likely to applaud than to give critical advice to the Labor Department.

The Department of HEW has also undergone an organizational shuffle under OSHA. The act required HEW to set up a National Institute of Occupational Safety and Health (NIOSH). The Department's response was simply to rename its Bureau of Occupational Safety and Health (BOSH). Attempts were made to move BOSH's laboratories from Cincinnati to Washington. Fort Detrick in Maryland, formerly used for chemical and biological warfare research, would have been an ideal site for the new institute, but the decision was made to remain in Cincinnati, with the director's office to be located in Washington within the Health Services and Mental Health Administration. Dr. Marcus Key, head of BOSH, was appointed Director of NIOSH. Again, a change in status without a change in attitude.

Appropriations: a test of Congressional support

To finance OSHA for the first six months of 1971 the Labor Department asked for a supplemental appropriation of $10.9 million, plus $100,000 for the Review Commission. The House Labor-HEW Appropriations Subcommittee at first denied this request, and then cut it by 30 percent. The amount allocated to the review commission was reduced to $75,000, with no provision for hearing examiners to adjudicate the large number of contested citations expected. The OSHA budget itself was cut by $3 million.

For the first full year of operation (fiscal year 1972), the Labor Department requested $31 million, a $24 million increase over past programs. Although this amount would have nearly tripled the existing federal safety budget, it would have averaged out to only forty-two cents per worker for the 57 million workers covered—which is less than the average currently spent by the states. HEW estimated that only $25 million was needed for their entire occupational health program. Yet according to their own study, the Frye Report, a basic program for worker health would require $50 million annually.[5]

The House Appropriations Committee approved these requests. But in the Senate, the Committee on Appropriations increased the budget, making a total of $34.9 million for the Labor Department. The Senate also increased HEW's budget.

The UAW submitted counterproposals to the Senate and House Appropriations Committees.[6] The UAW asked for $150 million for the first year of operations. At three dollars per worker, these funds could have provided an inspection staff of three thousand to carry out the Labor Department's mandate and also fund HEW's research. But Congress failed to heed the UAW's request and, for fiscal year 1972, budgeted $36.4 million for the Occupational Safety and Health Administration, and $25.1 million for the National Institute of Occupational Safety and Health.

The administration's fiscal 1973 proposed budget for OSHA once again sparked controversy. It allocated $67.5 million to the Labor Department, nearly doubling the 1972 figure, and $28.3 million to NIOSH. Most of the Labor Department's increase was slated for the continued development and operation of state plans. Over $6 million of the NIOSH budget was earmarked for research under the Coal Mine Health and Safety Act. A number of union officials attacked the budget in hearings before the House Appropriations Committee. One of them, comparing the proposed budgets for the Coal Mine Health and Safety Act, covering 200,000 workers, and OSHA, affecting 57 million workers, pointed out that the administration was planning to spend $310.95 per worker for enforcement under the Mine Safety Act, and $.41 per worker for enforcement under OSHA; $154.10 per worker for research under the Mine Safety Act, and $.50 per worker for research under OSHA.[7] The Senate Appropriations Committee was more responsive to the unions' pleas, adding $10 million to OSHA's budget, and $35 million to NIOSH's.

Implementation: HEW-NIOSH

Although renamed, HEW's branch of the safety and health agency has not yet cured the deficiencies that plagued its predecessor. Motion sickness seems to be the particular occupational health hazard of the new institute. Reorganized over and over again in previous years, NIOSH is now operating under the vague threat of another reorganization under consideration by President Nixon. Indeed, ever since the effective date of OSHA, NIOSH has never been sure what its status would be, thus delaying implementation of its responsibilities. Still, the familiar pattern of turtle-like progress and professional disdain of workers continues.

In order to assure publication of research and continuing progress in discovering the nature and exposure levels of toxic substances, OSHA required the Secretary of HEW to publish a list of such substances and their known exposure levels within six months of the date of enactment, or June 29, 1971, and at least annually thereafter. HEW failed to meet the statutory deadline, publishing a notice of the list's availability on July 23, 1971. The final list did not include the estimated fifteen thousand toxic substances in industrial use, but only about eight thousand—leaving the list of the remaining seven thousand substances to be published in the future.

NIOSH is in the process of developing so-called criteria packages for purposes of standard-setting. Literature searches and laboratory and epidemiological research are used to determine exposure levels and the necessary tests and labels that should be incorporated into a standard. The result is a criteria package, covering one toxic substance, that is then subjected to a lengthy in-house HEW review as well as consultation with outside experts. The first batch of criteria packages will cover only sixteen substances from the toxic-substance list, costing the institute from forty thousand to eighty thousand dollars apiece. NIOSH officials estimate that in two years they might be able to increase their productivity to twenty or thirty criteria packages a year. At this rate it will take over one thousand years to cover the basic list of toxic substances.

At present, NIOSH claims it takes nine months to produce one of these criteria packages. Yet according to Sheldon Samuels, an official of the AFL-CIO's Industrial Union Department who once worked for the National Air Pollution Control Administration, a criteria-setting agency, the process could be reduced to only a few weeks if an efficient data-collection system were used.[8]

As of September 1, 1972, the Institute had completed criteria packages for asbestos, coal dust, beryllium, noise, and carbon monoxide. Packages currently in progess involve arsenic, benzene, cotton dust, mercury, chromic acid mist, lead, ultraviolet rays, heat, cadmium, parathion, fibrous glass, and silica. NIOSH is also conducting industry-wide studies of the asbestos, beryllium, woodworking, printing, and coal tar industries, ferrous and nonferrous foundries, and the working conditions of operating engineers.

One way to pry criteria decisions out of HEW is to make a series of requests for emergency temporary standards under OSHA. This mechanism triggers a standard-setting procedure and would circumvent the long and unnecessary delay of criteria development. The AFL-CIO's Industrial Union Department has begun to experiment with this approach to jar NIOSH into action.

The act also provides for another, quicker evaluation of work hazards through requests for hazard evaluations. But NIOSH is still suffering from coordination problems, as well as from the difficulty of making hazards as reported by workers understandable to NIOSH scientists. A St. Louis Teamster local used OSHA to seek information about the chemical composition and hazards associated with materials in their workplace. All of the substances were labeled by trade name. The toxic-substance list contains generic names. While some of the labels contained health warnings, none listed the chemical formula necessary to ascertain the degree of hazard. Outside of HEW, the only source of this information is the individual manufacturer.

The union wrote to Mr. James Oser of the Scientific Reference Services Branch of NIOSH in July, 1971. He replied:

> In reference to your letter requesting chemical composition and health hazards with an extensive list of industrial trade products . . . I regret to inform you that at this time we do not have access to this type of information.

> We do hope that within one year we will have an
> operational file of material safety data sheets on chemical
> products. Your request should be directed to the manufac-
> turers as they have direct responsibility for providing health
> and safety information on their products.[9]

Not only does this letter reflect a total lack of concern for the
workers dealing with unknown chemical hazards, but it is an open
evasion based upon untruth. The American Conference of Government Industrial Hygienists, a group to which many NIOSH
people belong, has a trade-name index that would reveal the
chemical composition of these products. In addition, OSHA requires NIOSH to respond to requests for hazard evaluations
from workers. The union had committed the cardinal error of
writing to the wrong office and was getting a classic bureaucratic
shuffle. Fortunately the Teamster local in question was not as
ignorant as NIOSH assumed. They were working closely with the
Alliance for Labor Action's Summer Project on Safety and
Health. A member of the project contacted Dr. Moore, the associate director of NIOSH in Washington, and checked the law
for the proper request form. Based on this information the union
tried again, citing the provisions of OSHA. The request has been
referred to the Hazards Evaluation Services Branch of the Division of Technical Services, under Acting Chief Jerome Flesch.
Dr. Marcus Key, the director of NIOSH, once estimated that such
evaluations would take only a week following a NIOSH inspection,[10] but over two months passed before anything happened.
According to Art Buttons, a shop steward in the Teamster local,
"At first we didn't get any response, but when our local put some
heat on, we finally got reports back." Continued delay on the
part of HEW confirms the Department's past indifference to
occupational health as a low-priority item and permits continued
environmental pollution in the workplace.

Implementation—Labor Department

Voluntary compliance and education of careless workers are the
cornerstones of the Labor Department's program under OSHA.
As Assistant Secretary George C. Guenther noted, "The Training
and Education Office is looked upon as the key to the ultimate

success of the program. It will reach more people in a more effective way than the heavy club handed us by the act."[11]

In practice, Mr. Guenther's "heavy club" has come down as lightly as possible on violators. After the first ten months of operations, the Labor Department had proposed penalties totaling $1,711,995 for 75,864 violations, an average proposed fine of $23. At the Kawecki-Berylco plant in Hazleton, Pennsylvania, for example, a $600 fine was assessed for allowing exposure to beryllium dust to reach more than three times the allowable limit for eight hours. Twenty-eight nonserious violations resulted in a total fine of $328, including a $6 fine for permitting employees to eat and store food in an area where toxic materials were present. Four noise-level readings in four separate areas of up to 126 decibels, although substantially in excess of the permissible level of 90 decibels, provoked a fine of $75.

In reviewing fines assessed by compliance officers, hearing examiners of the Occupational Safety and Health Review Commission have ordered penalties to be increased in several cases. For example, after the Norfolk Shipbuilding and Drydock Corp. on three separate days permitted the use of non-explosion-proof lights in tank spaces not certified as "safe for fire," a compliance officer assessed a nonserious penalty of $175. The hearing examiner categorized the violation as serious and raised the fine to $1,000. The compliance area director in Philadelphia failed to assess any penalty when an employer did not report the hospitalization of an employee within forty-eight hours, as required by law. The hearing examiner, upon reviewing the case, assessed a penalty of $100.[12]

As a result of the Labor Department's emphasis upon education and compliance, all of the faults characteristic of the Department's prior safety activities are encrusting themselves upon OSHA's regulatory framework. But these shortcomings now have the potential to cause much more harm, since 57 million workers are covered by the new act.

Take, for example, the matter of standards. The act does not spell out any specific standards. Section 4(b)(2) provides that existing standards under the Walsh-Healey, Construction Safety, and National Foundation on the Arts and Humanities Acts "shall be deemed" occupational safety and health standards on the effective date of the law, April 28, 1971. According to this mandatory language, existing federal safety and health standards issued under these laws went into effect under OSHA as of April 28, 1971, and could have been enforced under the procedures set out in the new act. The Labor Department chose to ignore this provision because it felt that the legislative history and statutory language were unclear and would place too much of a burden upon employers.

Failure to invoke this plainly worded mandate exposed workers in non-Walsh-Healey plants to continued hazardous conditions, until the Department of Labor got around to establishing federal standards under Section 6(a) of the act.

The Department, however, avoided a legal challenge on this issue by swiftly promulgating an initial package of standards on May 29, 1971. (They acted so hastily that they published a set of Threshold Limit Values that were not in effect under Walsh-Healey and therefore were not federally established standards under the meaning of OSHA.) At an April 28, 1971, press conference, Undersecretary Laurence Silberman announced that over four hundred pages of national consensus standards and federally established standards had been sent to the *Federal Register* for printing under the authority of Section 6(a). By issuing the first standards under the authority of Section 6(a) the Department gave the misleading appearance of moving ahead at top speed. The law gave them two years to promulgate these standards, and they did it in four months.

If the Department had utilized Section 4(b)(2), existing federal standards would have immediately gone into effect. Invocation of Section 6(a) meant not only delay but an opportunity to promulgate national consensus standards as well as federal standards. The May 29 package included ANSI standards and standards adopted by the National Fire Protection Association, which the Secretary of Labor declared to be related to job health and safety. But no one at the Labor Department seemed to have read these standards carefully. They included a mass of petty detail hardly suitable for across-the-board imposition upon all workplaces covered by OSHA. For example, they spell out at excruciating length requirements for toilet facilities.[13] The indiscriminate promulgation of time-tested federal safety and health standards and consensus standards of dubious worth made the Labor Department vulnerable to attack by elements unsympathetic to OSHA, and as we shall describe later in this chapter, the onslaught was not long in coming.

In addition to ignoring the plain language of Section 4(b)(2), the standards issued on May 29, 1971, did not go into effect until August 27, 1971, and in some cases, until February 15, 1972. The purposes of this "familiarization period are to allow employers and employees to acquaint themselves with the standards, to encourage voluntary compliance, and to provide the states with additional opportunity to further consider and develop their plans to assume program responsibilities under the act."[14]

Delay might have been justified to allow familiarization with those trivial national consensus standards that the Labor Department ought never to have issued in the first place. But it was hardly necessary for existing federal standards and the more substantive consensus standards. As Congressman William Steiger commented in the conference committee report:

> The conference committee-reported bill provides for the promulgation of early standards by the Secretary of Labor without any hearing. However, these early standards would be limited to national consensus standards and established federal standards. These early standards would only be adopted pursuant to the authority in Section 6(a) within the first two years following the day of enactment, and when adopted would become effective immediately upon publication in the *Federal Register*.[15]

When questioned, Labor Department attorneys admitted that their action was inconsistent with a literal reading of the act, but defended the Secretary's decision on the ground that it was reasonable with respect to employers who would have to comply.

The Department partially acknowledged the error of its ways on August 13, 1971, by withdrawing the February 15, 1972, deadline for a substantial number of standards, including the Threshold Limit Values, noise, personal protective equipment, and hazardous materials. The official reason given was that "reconsideration of 'additional delay in effective date' provisions . . . has shown that the scope of the provisions is too broad and includes standards for which no delay is necessary for compliance."[16] However, some of the standards in the package were still delayed until February 15, 1972, and even longer, depending on the private consensus standard on which they were based.

The Labor Department's fetish for delay may specifically result in exposing workers to certain cancer-causing substances. The 1968 list of some four hundred TLV's contained the names of several contaminants found to be carcinogenic. However, the exposure limit, zero, was noted only in the appendix to the list. Regulations published by the Labor Department in 1969 subsequently incorporated the list into the Walsh-Healey Act.[17] The Labor Department now seems to be claiming that the appendix, with its zero tolerances, was *not* included within Walsh-Healey, despite the fact that the language of the regulations referred to the TLV's generally and made no exceptions. Therefore, the Department does not consider the zero tolerances for these carcinogens as federally established standards that would take effect immediately and without administrative hearings. This means that it will require a full hearing under OSHA to ban these cancer-causing substances. What makes this ironic is that during the Congressional hearings, one of these substances provided a convincing argument for federal regulation of the workplace. Betanaphthylamine, a dye ingredient linked to bladder cancer, had been banned by law in Pennsylvania, yet was being produced by a small company in Georgia[18]—a glaring example of the inadequacy of state job health laws.

OSHA is designed to protect the lives of all workers, whether organized or not. It is difficult to see how this goal can be achieved without worker participation and direction in the implementation of the act. Yet rules promulgated by the Labor Department have effectively stymied the creative use of the regulatory process to involve individual workers. Pressure from industry, and from labor unions who may feel threatened by worker independence, reinforces this position and locks the Labor Department into unimaginative and restrictive uses of its authority.

For example, Section 8(c)(1) of OSHA mandates that the Secretary of Labor issue regulations requiring "employers, through posting of notices or other appropriate means, [to] keep their employees informed of their protections and obligations under this act, including the provisions of applicable standards." Regulations implementing this section are now in effect. To carry out this notice requirement, the Labor Department has sent a single poster to each of the 4.1 million establishments listed by the Bureau of Labor Statistics. Employers may request additional copies. The poster has black print on a white background and thus will soon be smudged and obscured by factory grime and grit. It typifies the Labor Department's zest to involve and inform workers.

Also, out of loyalty to the myth of worker carelessness, the Department has squandered its funds on childish radio commercials such as:

> SINGER: Grab a sack of safety, let it fill your bag, workin' where you're careless is a hurtin' drag, yeah, take a sec and check around your own two feet, then sing a song of safety—it's a hap, happy beat, yeah!

And,

> CELEB: Hi. This is Joanne Worley. Don't get hurt on the job! Stay alert and practice job safety full time! For more information, contact the nearest office of the U.S. Department of Labor . . . Um Humm.[19]

No Labor Department commercial to date has sought to educate workers on their substantive rights under OSHA.

Especially important is the mandatory duty imposed on the Labor Department to issue regulations relating to injury statistics. The provision in the act negates the ineffective Z16.1 standard by including as "reportable injuries" all cases involving medical treatment, injuries resulting in loss of consciousness, regardless of the length of time and even if the worker returns to his job the same day, and injuries that restrict work or motion or require transfer to another job. Most companies have not kept track of total injuries, but under OSHA they must. A check of the records of some companies that have recorded total injuries confirms the need for this provision if reports are to be accurate. At the Cambridge Rubber Co. in Maryland, for example, there were 10 disabling injuries in 1970 and 531 total injuries. In the first month of 1971 there were no disabling injuries but 33 total injuries. The Photo Products Division of DuPont, in New Castle,

Delaware, recorded no disabling injuries in 1969 or the first 2 months of 1970. But 13 total injuries occurred during that period.

Employers are now required to keep several logs of injuries and illnesses and to post an annual summary for employee notice. These forms are to be made available to federal compliance officers during their inspections, but not to individual workers or the union. Several local unions are planning their own reporting system in an effort to get the same information. Most international unions still do not collect injury and illness figures on a country wide basis.

The reporting system has also attracted the interest of the Business Advisory Council on Federal Reports, an industry committee officially attached to the Office of Management and Budget in an advisory capacity. This Council makes recommendations dealing with all information request forms sent by the federal government to business. Jacob Clayman of the AFL-CIO's Industrial Union Department discussed the reaction of the Council to a proposed OSHA regulation on the reporting of injuries. He testified before a Senate subcommittee considering a bill to regulate advisory committees:

> Some weeks ago we received, from a source outside of the Office of Management and Budget, a notice of meeting of a panel of this august council on the subject of a log required [under OSHA]. . . . We were aware of the meeting, simply I suspect by accident. We had no formal notice. The trade union people as such are not invited, even though it was perceived the discussion would affect a vital area of our present and ongoing concern. . . .
>
> It is significant that the information to be gathered from the keeping of the log is not simply for the compilation of statistics for general public information. It is also a key part of the enforcement procedures under the Occupational Safety and Health Act. . . .
>
> Therefore, it is our considered judgment that the meeting may not be legally justified. Our understanding of the Federal Reports Act of 1942, the purported basis of the meeting, is that it was not intended to govern enforcement procedures. Therefore it does not apply to the proposed rule under the Occupational Safety and Health Act. . . .
>
> The meeting was another barrier, a way to hold up implementation of a rule essential to the safeguarding of life and limb in the workplace. To add insult to injury, the meet-

ing was held eight days after the deadline for public comments on the rule published in the *Federal Register*, which means that this great wisdom from this committee was accepted even subsequent to the date of the closing of the record to the general public.[20]

Section 17(f) of OSHA specifically prohibits advance notice of inspections, and provides a fine and/or imprisonment for anyone convicted of giving such notice without authorization from the Secretary. But at the same time, the act allows the Secretary of Labor to prescribe regulations dealing with inspections.[21] Taking advantage of this opening, the Secretary has proposed sweeping regulations permitting advance notice of inspections, thus mocking the law and the testimony of workers describing the abuses of advance notice.[22] In many instances companies will still be able to mount a temporary cleanup before the inspector arrives.

In his nomination testimony, Assistant Secretary Guenther could see only the reverse of this problem, and emphasized it as justification for advance notice of inspections:

> We also recognize, however, that there are a number of circumstances where it would not be administratively proper for us to dispatch inspectors to their various assignments, [or] compliance officers in their assignments, where they may wind up at a plant where there is an operation shut down. . . .[23]

The inspection of the Allied Chemical plant in Moundsville, West Virginia, provides an example of what really happens. The federal inspector inadvertently gave advance notice when he arrived at the wrong building. By the time he got to the correct plant, the company had shut down the mercury process that the inspector had come to examine. He had to return on another day to complete the inspection.

Mr. Guenther also argued that advance notice was desirable to make sure that employer and employee representatives are present for the inspection. Unions voiced strong objection to this argument, but the final regulations issued by the Labor Department oblige the employer, rather than the Department, to inform the employee representatives of the inspection date after the Labor Department notifies the company!

Additionally there is no guarantee that any given plant will be inspected. Based on the Labor Department's record under the Walsh-Healey Act, this is a tragic flaw. Without a mandatory inspection provision OSHA will never be adequately enforced, and the slaughter in the workplace will continue virtually unabated. With five hundred compliance officers to be hired by June, 1972, there will be one inspector for every 7,200 establishments.

The act does provide a partial solution to this problem, a provision permitting the Secretary of Labor to issue regulations requiring employer self-inspections. The United Auto Workers requested such a regulation, but the Department chose to ignore the request. Some unions are incorporating self-inspections into collective bargaining agreements. Section 4 of the United Steelworkers agreement with the National Can Co. provides:

> A regularly scheduled plant inspection shall be held once a month during working hours. Employees who attend such meetings during their regularly scheduled working hours shall not lose pay. A union member must be on the Inspection Committee.[24]

Another alternative, currently in operation, is the employee request for inspections. Priorities set up by the Department of Labor rely upon responding to complaints and requests for inspections and a limited general inspection program for certain target industries. Complaints received totalled 3,421 as of April, 1972, but that does not measure up to the deluge expected by the Department.

It was foreseeable that small businessmen would begin to complain about the expense of complying with the numerous standards now in effect under OSHA. Senator Carl Curtis of Nebraska has voiced his concern about the impact of the standards upon small firms. Letters of complaint to Senator Curtis, however, indicate that the problem is not that the Labor Department is now forcing these firms to comply with existing standards, but that small businessmen are apprehensive about the future costs of compliance. Since small businesses statistically have higher injury rates,[25] exemptions for them will defeat the purpose of OSHA. What they need is financial assistance for compliance expenditures. The Small Business Administration has funds available for safety and health expenses, but as of June, 1972, only three loans had been made for that purpose.[26]

In actual practice, there is some indication that the Labor Department has adopted a policy of selective enforcement and is not requiring immediate compliance with every provision of the act and the regulations. In a six-month evaluation of OSHA presented to Congress, Labor Secretary Hodgson reported that as of October 31, 1971, establishments were found to be in compliance in about 20 percent of the 7,918 inspections made. In another 20 percent, trivial, noncitable hazards were recorded. Thus, only 60 percent of the establishments inspected had violations. Under Walsh-Healey, the violation rate was estimated to be as high as 95 percent. A logical explanation for this difference is that under OSHA the Labor Department has not been citing all violations.

This has been confirmed by Assistant Secretary George Guenther, who admits that the Labor Department's regional and area offices are developing "checklists" of standards, and that in his view "nitpicking" safety regulations should be omitted from such lists. M. Chain Robbins, Guenther's Deputy Assistant Secretary, also confirmed the existence of this policy and added, "The proportion of establishments reported in compliance in September was approximately what it was in August, despite there being many more regulations in effect in September."[27] This was a tacit admission that the Department has set priorities for standards to be applied during inspections. But the Department has not considered these priorities important enough to make them public. Mr. Guenther remarked in an interview last August that "these are informal guidelines and as such I would imagine that there would not be all that much merit to publishing them."[28] Failure to disclose this inspection policy has also presented an inaccurate picture of how companies are complying with OSHA and can lead to even further mistrust of the Labor Department on the part of the unions and their members.

The Labor Department plans to direct its inspections, other than those provoked by worker complaints, at target industries on a "worst-first" basis, as measured by injury frequency rates. The target list includes lumber and woodworking products, transportation, meat and meat products, longshoring, and roofing and sheet metal. These industries, although recording a substantial number of work injuries, employ about 1 million workers out of a national workforce of 80 million, 57 million of whom are covered by OSHA. Also, the target-industries concept does not begin to face the occupational health problems prevalent in occupations with low injury frequency rates, such as the chemical, textile, rubber, plastic, and steel industries. Agricultural workers are totally overlooked, although they have the third highest injury rate and the highest death rate, and suffer a large number of job-related illnesses.

Assistant Secretary Guenther's explanation for the selection of the target industries is not very convincing. In addition to the injury frequency rate, he cites the consideration that a cross section of industries should be included—according to geographic location, size, and the existence of developed standards by which progress may be measured. The goal is to achieve the most good at the fastest rate. Thus, an agricultural program would be too expensive (and would not enjoy immediate, voluntary compliance by employers). The possibility of quick success for the program seems to be the paramount consideration. Further, Mr. Guenther admitted that the decision had a substantial political motive: the need to try to placate George Meany and organized labor.

Only four standards promulgated under OSHA deal with farm labor. They do not deal with tractor roll-overs or pesticide poisoning, although as of September, 1971, the Labor Department had promulgated standards for roll-over vertical protection devices for tractors used in the construction industry. The best guess is that the device would save six hundred farm lives per year, at a cost of $2.5 million. Great Britain, West Germany, Denmark, and Sweden have all managed to mandate this protection.

The only move in the direction of new standards has been the setting up of a task force within the Agriculture Department to study the idea of standards. Mr. Robert I. Gilden is the executive secretary of the task force. Its motto is: "Education-Engineering-Enforcement."[29] After a year of discussion with the Agriculture Department, the Department of Labor announced the formation of an Advisory Committee for Agricultural Standards. The committee includes no representatives of unionized farmworkers nor employees from the agricultural equipment industry.

The attempt to use the target-industry concept to head off criticism by the unions has failed completely. Attacks upon the program have been severe, especially from the labor unions. Belatedly, the Labor Department has announced a similar program for health hazards. At a NACOSH meeting on January 4, 1972, Mr. Guenther indicated that the Labor Department would concentrate on industries where asbestos, cotton dust, silica, lead, and carbon monoxide hazards were present. Noticeably missing from the list is occupational noise, which is estimated to affect 16 million workers.

The Labor Department has responded slowly to specific criticisms of its policies by spokesmen for workers. When asked whether OSHA marked the end of voluntarism in job safety and health, George Guenther replied, "We want industry's cooperation, and within the law, will do what we can to win it."[30] The Labor Department has not demonstrated any similar concern to foster the cooperation of organized labor or individual workers.

OSHA and the states

Passage of federal safety and health legislation could not have been achieved without the recognition that the states had failed in their traditional responsibility to safeguard the occupational environment. By providing for the setting of uniform national standards, developing a national system of workplace inspection, and allocating federal funds for research, OSHA furnishes the framework for a coordinated, all-out attack on the problem of work accidents and diseases.

Nonetheless, the states could not be excluded from the new program, particularly in view of their role under President Nixon's so-called new federalism. In addition, the federal record in the field of job safety and health was not so sparkling, especially under the Walsh-Healey Act, that an exclusively federal approach could be justified. Congress therefore decided to combine the resources available at both levels of government to combat in-plant pollution and industrial accidents.

Section 18 of OSHA allows the states to regain jurisdiction over safety and health standards provided they meet a list of requirements. The Labor Department is to give the states control only after they demonstrate their capability to meet these requirements. Unfortunately, the Labor Department has displayed a striking reluctance to exercise its ultimate authority over job safety and health. Instead, it has magnified the importance of the states' role out of all proportion to the intent of the act and engaged in a vigorous promotional campaign designed to surrender federal control at the earliest possible moment. Thus, on February 26, 1971, the Secretary of Labor sent to each state's Governor a letter spelling out the requirements for agreements under Section 18(h), which enables the states to continue enforcing their own work safety and health standards until either the Secretary of Labor approves a state plan submitted under Section 18(b), or until two years from the date of OSHA's enactment, whichever date is earlier. The letter indicates how swiftly the Labor Department can act if properly motivated:

> We are giving highest priority to the development of procedures which will enable the states to continue their current programs and to assume as quickly as possible the role envisioned for them by the act. The purpose of this letter is to provide you with an initial outline of our program plans in order that you may initiate such action as you believe necessary in your state.
>
> We contemplate issuing a package of basic safety and health standards on or immediately after April 28, 1971, the effective date of the act. As the act is written, these standards will preempt existing state standards and programs unless specific action is taken. Therefore, it is necessary quickly to enter into agreements under Section 18(h) of the act with states wishing to continue administration and enforcement of their standards pending approval of a state program plan under Section 18(b). Our overall objective is to have approved state programs under 18(b) in operation no later than July, 1972. This will necessitate the submission of state program plans by early 1972 in order that final action can be completed by May, 1972, and funding arranged by the start of this fiscal year.

This use of Section 18(h) to restore state administration and enforcement amounted to a blatant perversion of OSHA. Section 18(h) was never meant to authorize a broad re-delegation. It resulted from a last-minute amendment offered by Representative William B. Hathaway (D.-Me.) on November 24, 1970, the very day the House passed its version of the Act. Representative Hathaway made it very clear that the only purpose of Section 18(h) was to cover any hiatus between the effective date of a federal standard and the capability of the Labor Department to enforce the standard.[31] This would be accomplished by giving the Secretary of Labor *discretion* to let the states continue to enforce certain *specified* state standards, and *only during* this hiatus period. Thus, except for this narrowly delineated instance, the intent of Congress was to impose federal standards and federal enforcement until approval of state program plans under Section 18(b).

To date, forty-seven states, the District of Columbia, and Puerto Rico have signed Section 18(h) agreements with the Department of Labor. The basic commitment the state makes is to continue its current level of enforcement during this agreement period. To measure this level or establish a base line, the Department has set down a list of information that states should submit in order to get their agreements approved. This includes a description of the state standards, the designation of state agencies to handle enforcement, an explanation of the enforcement programs and procedures, and the dollars and man-years currently spent by the states for administration and enforcement.

One of the studies undertaken as part of the 1971 Alliance for Labor Action Summer Project on Occupational Safety and Health was to check several of the state agreements approved under Section 18(h) to see whether what was said on paper jibed with what was actually happening. Missouri and Massachusetts were chosen for these case studies.

The agreement filed by the state of Missouri with the Department of Labor listed nineteen safety and health inspectors with a budget of $244,000. But in the previous year, Missouri had a "director appointed by the Governor . . . an office staff of about twelve throughout the state, and twenty-four inspectors. . . ."[32] In addition to this cut in the number of inspectors, the Missouri budget had been reduced by 23 percent in 1970. Missouri law requires that every workplace be inspected twice yearly, so that in 1970 each inspector would have had to make seven inspections every day to meet the quota.

Even assuming that the requisite inspections were somehow made, there is still a question about enforcement of health and safety laws. Missouri believes in voluntary compliance by industry, rather than enforcement by sanctions. The description of the state's enforcement process in the Labor Department agreement states that "corrective measures are generally accomplished by education and voluntary compliance by the owner and/or operator of the facility involved."[33] A survey of the actual operation of the state's Division of Industrial Inspection, conducted by Daniel Berman of the Department of Political Science, Washington University, St. Louis, shows that:

> [I]n fact the punitive powers of the industrial inspection law have never been invoked. No one in the organization, including the director, could remember any fines or the sealing of any machinery under the provisions of the law. According to Director George Flexsenhar, enforcement would have to take place through the Attorney General's Office, a complex procedure that would divert resources from more pressing needs. The director does not believe that the Governor would back him up if he took a punitive attitude towards violators. He conceives of the problem of industrial safety as a problem of education rather than compulsion. Most violators eventually do what they should.[34]

Corrections are made in response to work orders with which the owner must comply within ten days, or be subject to fines and a plant shut-down. This authority, as noted above, has never been exercised, nor is the ten-day limit observed. A rigid policy of voluntary compliance abhors all deadlines. A special inspection of the St. Joseph Lead Co. in April, 1968, found several violations that resulted in the issuance of a work order. The company received ninety days in which to comply, but after six months it had not corrected a single violation. A new work order was issued, with no penalty and an additional ninety days to comply.

State occupational health laws also feature feeble or nonexistent enforcement. Under Missouri law employers must give a monthly physical examination to workers exposed to occupational hazards, and the physician is supposed to report any illness or disease within twenty-four hours to the Division of Health. What actually happens is described by the director of the Environmental Health Section of the Missouri Department of Public Health and Welfare, in a letter to Daniel Berman:

Due to the infrequent receipt of a report of occupational disease, the Division of Health has not prepared a statistical summary. Such a summary would have little or no significance. The Division of Health has not taken action to obtain enforcement of the provision of this section [penalty on the physician for not reporting]. You will note that the section refers to a requirement that employees exposed to certain materials be examined every thirty days. We interpret this section as requiring a report of all examinations be submitted to the Division of Health whether or not an occupational disease is found or diagnosed. We do not have the staff to process the thousands of reports that would be received each month, nor do we believe that there is sufficient public health significance to justify the additional funds that would be required.[35]

In other words, the agency doesn't collect the reports as the law requires. It seems foolish to expect any change in Missouri under OSHA. The Governor has designated an Occupational Safety and Health Task Force to advise him in this area. The chairman is George Gorbell, of Monsanto in St. Louis, whose testimony before the Senate Committee on Labor and Public Welfare in 1970, in support of the Nixon administration bill, stressed the importance of "voluntary safety endeavors of industry and of professional groups in the United States."[36]

Massachusetts has submitted a full state plan instead of the temporary two-year agreement under Section 18(h). While the Labor Department cannot approve this plan until Massachusetts complies with final regulations, the plan merits scrutiny as an example of how one state initially tackled the problems of meeting federal criteria.

The Massachusetts Department of Labor and Industry will issue standards subject to review by a labor-management board. These standards will take the form of industrial bulletins covering hazards from anthrax to window cleaning. The Massachusetts plan asserts that these bulletins already meet the requirements of Section 6(b)(7) of OSHA, under which standards must include warnings of hazards, relevant symptoms and emergency treatment, specifications for protective equipment, medical examinations, and monitoring. This is a remarkable achievement, since the Labor Department hasn't issued any specific standards under Section 6(b)(7) yet. One hopes, perhaps in vain, that the Labor Department will not accept any such claim on the part of the states that their laws meet federal requirements that do not exist.

The main failure of the Massachusetts plan to meet federal law is in the area of employee rights. The plan says that Massachusetts inspectors "welcome employer and employee representatives on tours of the plants." Of course, a welcome is not the same as a right, for welcomes wear out. Section 8(e) of OSHA guarantees walk-around rights to the workers, and Section 8(c)(3) entitles them to observe monitoring of workplace hazards. Urban Planning Aid, an Office-of-Economic-Opportunity-funded project in Boston, analyzed the Massachusetts state plan submitted to the Department of Labor. They reported that, under state procedures, an employee representative does not accompany the inspector or observe monitoring unless the *company* requests it. This practice clearly does not meet federal standards.

The Massachusetts plan also claims that some of the bulletins issued by the Department of Labor and Industry meet the requirements of OSHA's Section 8(c)(3), which deals with the recording of employee exposures to toxic substances. Yet the plan makes no mention of another requirement spelled out in Section 8(c)(3), that employees have access to their own records. No Massachusetts laws or rules guarantee workers access to records (unless they subpoena them for a court case). In conversations with state officials, representatives of Urban Planning Aid found out that, under existing state policy, workers can look at records of inspections, tests, or exposures if they come down to the state offices. But they are not allowed to take them away or make copies. State practices, therefore, discourage the dissemination of job safety and health information to the workers.

Section 9(b) of OSHA requires that citations issued under the act be posted at or near the place of violation and list the time within which corrections must be made. Section 8(c)(1) provides that by posting notices, or by other means, employers must keep employees informed of company obligations and the protection afforded them by the various applicable standards. The Massachusetts plan contains no procedure for posting written orders. Indeed, state inspectors often issue oral, on-the-spot orders for corrections, with the result that employees have no way of knowing which standards are supposed to protect them or how these standards are being applied.

Under Section 8(f)(2), employees are entitled (on request) to a written explanation if a citation is not issued for a violation they assert exists. The Massachusetts plan says that the issuance or nonissuance of a written order is reviewed as a matter of departmental procedure, but it does not mention participation by employees in this informal review.

As in Missouri, enforcement is nonexistent. Health and safety efforts reflect a total commitment to education and training. Section 17, containing the basic penalty provisions of OSHA, requires fines of up to one thousand dollars for serious violations of the act. The Massachusetts practice is to issue a written order for a serious violation and then to conduct a reinspection to see whether the violation has been corrected. According to state officials, no penalties for health violations have ever been assessed. Urban Planning Aid has reported that in 1969 the state issued 122 written orders for health violations and prosecuted none. This lack of enforcement destroys any initiative for self-policing on the part of employers. If an inspector finds a serious violation, the employer can correct it at his leisure, without fear of being penalized. So why worry about job safety and health?

Neither Missouri nor Massachusetts lives up to its claims to the Labor Department in the Section 18(h) agreements. There is little reason to suspect that the reality behind the agreements made by the other forty-seven jurisdictions that have filed them is much different. To date, the Labor Department has not attempted to cope with this situation or even admit its existence. The act does permit scrutiny of operating state programs for at least three years before the Secretary of Labor relinquishes his enforcement authority. But the Secretary is not legally required to do this, so that he is free to approve a plan and never look to see what the state is actually doing.

Richard Wilson, an official in the Labor Department's Office of State Plans, has indicated that the states' enforcement records will have some bearing upon whether a proposed plan is accepted.[37] For example, if a state claims no need for sanctions similar to those specified by OSHA but proposes reliance upon a red-tag shutdown procedure, then the history of the use of this procedure in the state will be important in assessing whether the state plan is "at least as effective" as the federal one. The Labor Department's regulations, however, do not require the states to submit data on the actual enforcement of job safety and health laws and regulations in the states. The Department seems to be assuming that the states will provide this information, which would be contained in court records and inspection reports.

Thus, if the proposed state plan contains sanctions similar to those spelled out in OSHA, the state does not have to supply any enforcement data, and the Secretary of Labor need not ask for it. If different sanctions are proposed, the states are still not required to supply enforcement data, but the Labor Department expects them to produce it. In either instance, Labor Department officials remain unwilling to take an uncompromising approach toward enforcement and thereby continue to rely upon the policy of voluntary compliance.

In addition to its failure to survey the enforcement practices of the states, the Labor Department has sought to undermine the substantive rights OSHA grants to workers. During a surprise visit to the second NACOSH meeting in September, 1971, Undersecretary Silberman announced that it would exceed the "legal and policy guidelines of the Department to require specific federal rights in the state plans," unless these rights were specified in Section 18, the state-plan section of OSHA.[38] He used as an example one of the most important rights guaranteed to workers by OSHA—the walk-around inspection.

Section 8(e) requires that an employee representative accompany the federal inspector on an inspection of the workplace. Section 18(c)(3) allows the Secretary of Labor to approve a state plan containing inspection provisions "at least as effective as that provided in Section 8." What Mr. Silberman was doing was telling the states that their plans might be approved even if they did not contain a walk-around inspection provision. The Labor Department's proposed regulations classified the walk-around as merely one way the state can authorize employees to notify inspectors of possible violations in the plant.

To date the Department has not furnished any legal authority for its "legal and policy guidelines." But what it has done is to open up a way for the states, in collaboration with the Labor

Department, to extinguish many of the specific requirements of OSHA by making them "examples" for the states to follow or modify as they please. Thus, the Labor Department has managed to convert the seemingly innocuous state-plan section of OSHA into a trap for the destruction of worker rights.

During the legislative struggle over OSHA, the U. S. Chamber of Commerce circulated to its lobbyists a memorandum decrying the fact that the phrase "at least as effective as" in the Daniels bill meant that the state plans had to be at least as tough as, or tougher than, the federal standards. Therefore, the Chamber argued, every effort should be made to secure the adoption of the provision in the Javits-Ayres bill, which would allow the states to administer their own plans "if substantially as effective" as the federal standards.[39] OSHA, as passed, contained the language of the Daniels bill on this point. Yet the thrust of Silberman's statement was that the Labor Department would administer OSHA as though the provision in the Javits-Ayres bill had been included—an unbelievable display of disdain for the plain language and legislative history of OSHA (and no doubt heartwarming to the U. S. Chamber of Commerce).

It is very difficult to visualize an alternative that would be "at least as effective" as the walk-around. Mr. Wilson, in the Labor Department's Office of State Plans, declares that he has found the states willing to include a provision for a walk-around in their plans. Indeed, Wisconsin has already implemented such a provision. But problems may arise over the issue of whether third-party personnel may accompany the inspector, or whether worker representatives will be paid for time spent on the inspections. Assistant Secretary Guenther has already made a ruling on the latter issue. On March 2, 1972, he rejected a union complaint and decided that the Mobil Oil Co. in Paulsboro, New Jersey, did not discriminate against its employees by refusing to pay them for time they spent on a walk-around inspection. The workers lost over four hundred dollars for their exercise of a right guaranteed by OSHA.

In any event, labor representatives at the NACOSH meeting strenuously objected to Silberman's announcement. But in a subsequent meeting with Silberman and Guenther, George Taylor of the AFL-CIO and Jack Sheehan of the United Steelworkers inadvertently ended up making things even worse. They pointed out an inconsistency in the Labor Department's proposed regulations, which used the walk-around as an *example* of how employees could notify inspectors of violations, but *required* state plans to include authority to promulgate emergency temporary standards as a means of protecting workers from unexpected dangers. The labor union officials were arguing that the walk-around should also be mandatory, but it turned out that they gave the Labor Department officials a pretext to weaken the regulations even further. The Department now interprets the authority to promulgate emergency standards as an *example* of how the states might protect workers from new and unforeseen dangers, rather than as a *requirement*.

The Labor Department policy on permanent state plans also drew fire from Taylor and Sheehan. Under the "developmental plan" concept, the Department would approve a state plan and allow it to go into effect, with the stipulation that the plan must meet all the requirements of OSHA in three years. The union men argued that OSHA preempted the states' authority and required the Labor Department to maintain its control over the workplace until the states actually met all the requirements of the federal act. But OSHA proved to contain a wide loophole on this point, and the Labor Department was characteristically quick to take advantage of it. Section 18(c) allows the Secretary to approve plans which either do or *will* meet the requisite standards.

The evidence from Missouri and Massachusetts indicates the wide gap between word and deed in the agreements for interim state control of the work environment. The gap between promise and performance in the permanent state plans may well be much wider.

Although forty-nine states have filed notices of intent indicating their willingness to submit a plan within nine months of the notice, it is by no means certain that the majority of these states will in fact submit permanent plans. Barry Brown, director of Michigan's Department of Labor and a member of NACOSH, estimates that "there will be approximately fifteen states which will end up with comprehensive state plans."[40] As of October 1,

1972, nineteen states had submitted developmental plans for consideration by the Labor Department.

There are several obstacles that militate against state participation. Many states, for example, would have to make legislative changes in order to include agricultural workers under safety and health standards, give an inspector a right of entry into workplaces, and authorize the necessary personnel to continue effective programs. States also need mechanisms for developing, enforcing, and hearing challenges to safety and health standards. A more basic stumbling block is money. The act only provides for grants on a fifty-fifty basis for the actual administration of state programs. The higher the cost of a comprehensive plan, the less likely it is that the states will be willing or able to match it with their own funds. Failure of the states to make these improvements or raise the necessary money will result in federal control.

Over the next several years the Department of Labor will implement its program of state control. Amendments to OSHA may remove the financial barrier to state control by making the grants ninety-ten for the administration of state programs, but the unions and other interested groups will have only themselves to blame if the Labor Department approves state plans that meet the on-paper requirements but in fact are not implemented. The Department will not compile information on nonimplementation for it does not want to acknowledge its existence. Labor must make it crystal clear that the unions will not sit on their hands but will dig up the records of the states, disclose their activities to the public, and demand that the Labor Department carry out its responsibilities.

Since 1972 was an election year, it is not surprising that organized labor's attacks upon the state-plan regulations were impugned as politically motivated. Barry Brown, for example, dismissed the unions' criticisms as "designed solely to embarrass the present administration on all of their labor involvements," and stressed what he sees as a contradiction between union hostility to the Labor Department in Washington and union support for state programs.[41] But labor's disenchantment with the federal administration of OSHA has been simmering ever since

the Labor Department began to implement the act and therefore cannot be categorized as a posture adopted for immediate,
short-range political purposes. The AFL-CIO has unsuccessfully
requested an administrative hearing on the state-plan regulations issued by the Labor Department. Having lost that battle, the
federation has developed a model state law and is planning to
lobby for it in a number of states. The model law incorporates
many of the guarantees specified in OSHA and in some instances
goes beyond the federal law. (For example, it contains no provision for advance notice of inspections.)

The performance of the unions

The Labor Department's zeal to pass responsibility for job safety
and health back to the same state officials whose shabby performance had created the need for a federal statute underscores
the importance of organized labor's role as defender and partisan
of OSHA. The crucial question is whether the unions will abandon rhetoric and paper-shuffling and take advantage of the opportunities created by OSHA and the growing health and safety
consciousness of workers. In the months following the enactment
of the new law, the unions have had enough time to formulate
the beginnings of their response to OSHA and the way it is being
administered. The United Auto Workers (UAW), with 1.3 million members, and the Oil, Chemical and Atomic Workers Union
(OCAW), with 165,000 members, provide graphic examples of
contrasting approaches to job safety and health.

The basic document setting out the UAW policy toward
OSHA is an Administrative Letter sent by UAW President
Leonard Woodcock on July 19, 1971, to all UAW local union
presidents. It reads like a set of Labor Department regulations
and is just as inspirational. The UAW position is to call for a
"chain-of-command" procedure for handling safety and health
complaints. Although action at the local level is not expressly
forbidden, the locals are urged to refer complaints to the union's
regional service representative, who would then decide whether
or not to request an inspection. According to Lloyd Utter, safety
engineer and one-fourth of the UAW's safety and health staff,

"Without this chain of command, we would run the risk of having self-appointed experts in hundreds of plants operating singly, on the one hand, or of being inundated with requests here at head-quarters in Detroit, on the other."[42]

Neither risk has materialized, but not because of the success of the Administrative Letter. Indeed, in the five months following the distribution of the letter, no one has followed the chain-of-command procedure either. As of November, 1971, the Social Security Department of the UAW (which took over health and safety activities in March, 1971) was aware of only two complaints filed with the Labor Department by UAW locals. Both of them were initiated by the rank and file, without regard for the strictures of the Administrative Letter.[43]

Caution is the watchword of the UAW approach. But caution, combined with the problems inherent in the union's safety bureaucracy, succeeds only in anesthetizing worker initiative.

Another difficulty in relying on the chain-of-command procedure is that it puts more work onto the already overburdened regional service representatives in the UAW's eighteen regions. Because of UAW staff cuts, the service representatives are older men and women with high seniority. They have great working knowledge of plant conditions and the political situations in their locals, but they are not physically capable of the massive effort a successful safety and health program demands. Educational efforts aimed at the regions in two-day briefings on the law and the Administrative Letter will rarely filter down to local unions because of time constraints on service representatives. The UAW does plan to hold educational courses at Black Lake, the Walter Reuther Educational Center in Michigan. But sending local union safety representatives to the wilds of Michigan for training sessions may not be as effective as sessions conducted near the workplace.

UAW President Woodcock has come out strongly for a vigorous implementation of OSHA. In a speech before the Engineering Society of Detroit, he said the law

> . . . is the best hope they [57 million workers] have ever had that they will be able to earn a living with minimal risk to life and limb. And to the extent that the Nixon administration loses sight of that hope and need, and comes to

think of the Occupational Safety and Health Act as a show-case or promotion stunt camouflaging the same old health-safety runaround, there is going to be hell to pay.[44]

Yet the same advice can be given to the UAW itself.

Some people in the UAW are aware of the need to cut through their own red tape. As one service representative said at a regional briefing in New Jersey, "The law means nothing. When a machine is unsafe, you just shut it down." It is hard to believe that these people do not realize what it means to see the National Safety Council emblem emblazoned on the copy of the law distributed by the UAW, or the extent of the harm inherent in some of the model collective bargaining clauses suggested by Detroit for local negotiations. Only a union safety department that has bought—or fallen for—the industry line could suggest acceptance of the American National Standards Institute and National Fire Protection Association Codes as minimum safety conditions. Even if OSHA enacts them into law, the UAW doesn't have to put them in their contracts. Cynicism is the only appropriate response to Section XIII of the suggested model clauses: "The company and the union each agree, as a demonstration of their sincerity in this Occupational Safety and Health Program, to maintain membership in the National Safety Council."[45] Lloyd Utter has said, "We and corporate safety men are usually headed in the same direction."[46]

The UAW *Washington Report*, a weekly newsletter, is a consistent outlet for information on occupational safety and health. Very readable and often graphic when describing effects of in-plant pollution on workers, it portrays the UAW as a champion of worker safety and health. But a closer look reveals that the prime source of this preoccupation is the report's managing editor, Frank Wallick. Not an official member of the safety and health bureaucracy of the UAW, Mr. Wallick is an effective spokesman for the right and ability of working men to control their own working conditions, and he enjoys the support of several active regional and local UAW officials. He was one of the motivating forces in setting up an occupational safety and health project with the Alliance for Labor Action Summer Intern Program. The Alliance, a joint effort of the Teamsters and the UAW, worked in the area of social action and organizing. It initiated

Project Workplace Survival, which developed manuals for shop stewards telling how and when to call for federal inspections and what types of hazards existed in auto plants. These manuals also included suggestions for monitoring equipment and descriptions of hazard symptoms. Field projects in Boston and St. Louis developed rank-and-file and community-supported safety and health operations. In addition, members of the project collected the experiences of other local unions and compiled a booklet on tactics. Another member did legal research on proposed OSHA regulations, so that the UAW could submit comments to the Labor Department. To date, only the legal comments have been actually used. The other project reports run counter to the Administrative Letter, since they are aimed at union locals and not at the regions. According to George W. Strugs, Jr., a consultant in the UAW's Social Security Department, "It is my understanding that the law booklet, in its present form, invites and encourages individual action; this, of course, conflicts with our policy. . . ."[47] The reports are now gathering dust on a shelf in UAW headquarters in Detroit.

The Oil, Chemical and Atomic Workers Union (OCAW), with headquarters in Denver, Colorado, outlined a procedure similar to the UAW's chain-of-command in a letter from President A.F. Grospiron to the locals in June, 1971. However, according to union policy, "Local unions are free to make direct requests to the Department of Labor for health and safety inspections, or such requests may be made by an international representative on behalf of the local."[48]

Part of the difference between the two unions may lie in the nature of their membership. Close to half of the UAW members are in one state, Michigan, inevitably leading to more centralization, while the OCAW is spread out across the country, with small numbers in very large plants. Another crucial difference is the OCAW's lack of a safety bureaucracy. The union's resources include Fred Linde, a chemical engineer in Denver, and the staff in the Washington Legislative Department, under Anthony Mazzocchi. The program Mr. Mazzocchi has developed is essentially "run on people, not money," and so far it has produced results.

Because inspections are sporadic happenings, the OCAW

focuses upon bargaining and grievance procedures. These tactics require information and knowledge from the local members. To that end, Mr. Mazzocchi and his assistant, Steve Wodka, are developing the concept of "plant situation teams." According to Mr. Mazzocchi, the objective is to train "local members themselves so they can control their own work environment through procedures there—bargaining and grievance, rather than inspection—but also to be able to check up on the inspector."[49]

District 8 of the OCAW, centered in New Jersey, is the focal point for this pilot project. The problems in the New Jersey plants were described by local members in a series of "consciousness-raising sessions" held while OSHA was still before Congress. The OCAW has devoted its scant resources to utilizing this information through a fourteen-week training course at Rutgers University in New Jersey, conducted by the Scientists Committee on Occupational Health, a group of volunteers organized by the OCAW. Classes contain about fifty-five union members, including some participants from the UAW, who will soon receive their own courses from the UAW's Region 9 in New Jersey. The National Institute of Occupational Safety and Health lends equipment for demonstrations. The course is designed to produce "barefoot doctors"—workers who will know what carbon tetrachloride can do to their livers and kidneys, what substitutes are available, and what hazards these substitutes entail. As Dr. Jeanne Stellman and Dr. Susan Daum of the Scientists Committee have noted, "It doesn't demand a professional to take the air samples [in a plant]. Workers can be trained to do these things by scientists who are willing and able to overcome the aura of mystery surrounding their professions. Workers must in turn make sure that the professionals know what the problems of the workplace are."[50]

The OCAW has also been joining with other unions to take advantage of services offered by the Medical Committee on Human Rights, which has chapters throughout the country. At a meeting in Chicago several months ago, the UAW regional office and the Medical Committee sponsored an organizing session for some twelve UAW locals, along with representatives from the Steelworkers, Teamsters, Meat Cutters, United Electrical Workers, OCAW, and several independent rank-and-file groups.[51]

Commented Charles Hayes of the Meat Cutters, "We've had company doctors; maybe we can get some union doctors."[52]

Anthony Mazzocchi's official title with the OCAW is citizenship-legislative director, and hence much of his safety and health work is basically a collateral duty. His big problem is money. The OCAW cannot continue to ask local members to assess themselves for the cost of medical tests, as was originally done at the Kawecki-Berylco plant in Hazleton, Pennsylvania. (The company eventually agreed to defray 50 percent of the cost.) The OCAW people have begun to face up to their financial difficulties. They are developing a laboratory subscription system to cut the cost of sending medical tests to commercial laboratories. Members of UAW locals in Michigan also have the opportunity to trim occupational health expenses by participating in Health Watchers, Inc., a union-sponsored, computerized diagnostic clinic that reduces hospital costs for a battery of tests from five hundred dollars to fifty dollars.[53] Any group medical program can similarly be directed toward occupational health goals and thus reduce reliance upon company doctors.

Another union that is beginning a job safety and health program with an orientation similar to that of the OCAW is the United Electrical Workers (UE). Its 165,000 members are exposed to hazards from extremes of temperature, noise, chemicals, fumes, and asbestos. According to Larry Rubin, a staff writer for the UE's newspaper, "Workers are aware and frightened of being injured and made ill by their jobs."[54] He sees the union's role as one of channeling worker concern and finds that more is accomplished when the workers take the initiative. The UE program is highly decentralized. "We are against an elitist kind of approach," says Mr. Rubin. Therefore, the UE trains its organizers to stress safety and health in their organizing activities. In addition, at the international level, the union encourages the participation of the president and one or two others from each local. But the ultimate decisions on job safety and health issues are left to the workers, since decisions and priorities fashioned by outsiders will inevitably meet with resistance from the rank and file. The UE is beginning to build relationships with local doctors, medical students, and professionals.

Other member unions of the AFL-CIO are experimenting

with different approaches to job health and safety. The International Printing Pressmen and Assistants Union of North America, the International Association of Heat and Frost Insulators and Asbestos Workers, in addition to a number of other international unions, are cooperating with the American Cancer Society on a $1 million, year-long study of their members to detect cancer-causing agents' in the workplace. The Society's press release quoted Dr. Irving Selikoff of the Mt. Sinai Environmental Sciences Laboratory and Dr. Cuyler Hammond, the Society's vice-president for epidemiology and statistics, as saying that:

> Priority will be given to groups exposed to agents now under suspicion as possible carcinogenic or co-carcinogenic agents . . . and agents to which the general population is exposed. While seeking the effects of occupational exposure to various agents, the study will also determine whether the same agents result in an increase in cancer rates among people with non-occupational exposure.[55]

It is important that the unions see to it that the American Cancer Society follows up any findings with special efforts to remove carcinogenic agents from the workplace, so that the workers who have served as subjects of the study will derive direct and immediate benefit from it.

In contrast to these activities by the smaller unions, the powerful International Brotherhood of Teamsters has displayed little interest in OSHA or its implementation. The Teamsters, with no Washington staff assigned to job health and safety, neither lobby nor participate in Labor Department proceedings relating to hazards that affect union members. Only with the appointment of Robert Kasen as editor of *Focus*, the Teamsters' weekly newsletter, have things begun to stir. Kasen has inaugurated a series of articles designed to educate the rank and file about job hazards and to spread information about local union projects, such as the "small-budget approach" to on-the-job health taken by a Teamster local in St. Louis. In addition, Kasen has printed and publicized a pocket-sized booklet, *Job Health, Safety, and You* (the same materials developed by the Alliance for Labor Action in the summer of 1970 but never used by the UAW). Over fifteen thousand requests for this eminently readable booklet have been received from the locals.

Interference with management's conduct of its business is held not to be a proper subject for union negotiations. For example, under existing law, a union could not bind General Motors in negotiations to build a nonpolluting car. But to what extent are in-plant safety and health issues subject to this restriction? Some courts have held that safety and health concerns fall within the definition of conditions of employment so that workers cannot be discharged for protesting job hazards.[56] But do unions also have the right to direct how management can eliminate safety and health problems by negotiating clauses describing equipment or requesting installation of monitoring devices—in effect, helping to design the production process? As of now, debate on this point is more theoretical than practical. The unions are just now beginning safety and health negotiations and are focusing on setting up procedures allowing access to plants and inspection reports, and guaranteeing the pay of workers assigned to safety and health duties. This was part of the goal of the United Steelworkers of America in their 1971 contract negotiations with the can, aluminum, basic steel, and nonferrous metal industries. At the Second International Conference on Safety and Health held by the Steelworkers in March, 1971, attention focused on OSHA and the collective bargaining process. One union official declared:

> In the past the industries with whom we bargained have long known that the United Steelworkers of America was not serious on safety, merely using safety as a window dressing, getting concessions on other issues. It is no wonder that we have made little progress. This is a true statement and you know it is.[57]

A change in the image of the Steelworkers began in 1971, when the union set out to obtain,

> . . . as a fundamental part of the protections that are ours by right, a clear, unequivocal privilege on the part of each and every member of the United Steelworkers of America to refuse without penalty to work under unsafe and unhealthy conditions.[58]

"Without penalty" means with the right to pay or transfer to another job for time lost when the regular job is unsafe. It does not simply mean protection from discharge if a worker walks off an unsafe job. Another area included in bargaining was the definition of the functions of joint safety and health committees, including payment for time spent fulfilling safety and health duties.

The Steelworkers enjoyed varying degrees of success in their safety and health bargaining endeavors. The right to refuse to work without penalty was granted in the Bethlehem Steel, Armco Steel, Anaconda, and American Smelting and Refining Co. agreements, but to avoid a penalty the employee has to be correct about the existence of "an unsafe condition, changed from the normal hazards inherent in the operation."

Payment for safety and health duties is an essential concomitant to the right of employee representatives to accompany inspectors. Union members will be reluctant to press for inspections and meetings if they will be docked for the time spent on them. None of the Steelworkers' contracts deals with the employee representative specifically, although the Anaconda agreement has the broadest language:

> Section 2(f): Union Joint Health and Safety Committee members shall suffer no loss in wages while on official business in connection with their duties.[59]

Some agreements, mainly in the steel industry, specify that safety and health committee members are *not* to be paid, even for meetings held during work hours.

General Electric's policy is not to pay for walk-around, according to Thomas Hilbert, the company's labor relations counsel.[60] At the Continental Oil Co., workers are paid for walk-around during their shift, but not if the walk-around goes into overtime.[61]

Protection for employees working with toxic substances developed in the aluminum industry bargaining. The Alcoa and Olin language is as follows:

> When an employee is temporarily reassigned to another department as a result of the medical department's determination that his exposure to a toxic substance calls for such temporary reassignment, he shall receive for hours worked his regular rate of pay or the pay of the job classification or job classifications to which he is assigned, *whichever is higher*, for a period of thirty (30) days following reassignment, or upon return to his former department, whichever

is sooner. The local parties may mutually agree to extensions
of the rate retention period. [Emphasis added.][62]

Kaiser's agreement omits the phrase "whichever is higher," and
Reynolds did not include the language in the contract, but rather
in a letter of understanding.

Contracts are like laws governing relations between the
company and its employees. Like laws, their success in benefiting
employees depends on how well they are implemented. As Ben
Fischer, director of contract administration for the Steelworkers,
reminded union delegates at the 1971 conference:

> Having developed it, having negotiated it, having it put in
> legally binding contracts, how do we then give these words
> meaning in the daily lives of our members in their work-
> place, because everyone in this room knows all too well that
> it is one thing to write a contract and negotiate it—it is
> quite another thing to make it meaningful in the day-to-day
> application and in the day-to-day problems that our mem-
> bers face.[63]

Like other unions, the Steelworkers have OSHA educational and
training programs in operation and on the drawing board. In conjunction with these programs, locals under the new contracts need
assistance in taking advantage of these provisions, expanding
their interpretation, and exercising their rights to safe and healthy
working conditions. Strong contract language is a necessary
corollary to any implementation of OSHA and must be centered
on individual locals, not the international bureaucracy. Contracts
are tools for the working man and woman, enabling them to
exercise control of the quality of their lives in the workplaces.

Grievance procedures provide methods for putting these
tools into action. If well designed, they can deal quickly and effi-
ciently with many job safety and health complaints. The Amal-
gamated Clothing Workers (ACW), whose members risk respiratory ailments, dermatitis, and eye irritation from exposure to
formaldehyde and chemically treated synthetic fibers, is encouraging the use of labor-management committees to implement
OSHA. William Elkuss of the ACW's Education Department
points to the low priority assigned by the Labor Department to
hazards in the clothing industry as the main reason for this approach: "Given the magnitude of the asbestos, coal mine, and

textile problems, little will be done for our members, so we decided to take the initiative." Locals in the ACW's St. Louis Region have developed an eight-point agreement to handle safety and health complaints via labor-management committees. The steps include the gathering of hazard data on fabrics and chemicals, and the use of experts and laboratories to investigate health problems.

While each union strikes out on its own in interpreting OSHA and setting safety and health priorities, the AFL-CIO coordinates overall policy *vis-à-vis* Congress, the Department of Labor, and HEW. George Taylor is executive secretary of the AFL-CIO Standing Committee on Occupational Safety and Health, which is only one of his several jobs in the safety, health, and environmental fields. Labor's efforts have been only partially successful, and in Mr. Taylor's view, dealing with the Labor Department is not very useful. The unions object to what they see as the pro-management policy of the Department, ranging from appointments such as that of Boeing official M. Chain Robbins as Deputy Administrative Assistant, to decisions on regulations, especially for state plans. The Department's actions have resulted in a union boycott of its training sessions since the Labor Department gave Boeing the contract for conducting the training sessions despite assurances to union officials that Boeing would not be awarded the program.[64]

The AFL-CIO has been highly vocal in its criticism of the Labor Department. "We can't enforce, so we can complain," says George Taylor, expressing the basic tactic of the AFL-CIO. Such methods may work against a Republican administration, but when a member of the Study Group asked Mr. Taylor what would happen if a Democrat were elected to the White House, he stated, "The Democratic administration would be more difficult, and complaints would have to be under the table, since Democrats are our boys." He said he wished he could just "sit back, light up a cigar, and say, 'They're going to do it.' "[65]

Another part of the AFL-CIO's program stems from the Industrial Union Department (IUD), headed by I. W. Abel of the Steelworkers. Under Sheldon Samuels, who has the help of one student intern, the Occupational Health and Safety and Environmental Affairs Department of IUD focuses on training and standards, and comments on Labor Department policy. Training institutes run by the AFL-CIO Labor Studies Center at eleven universities around the country seek to certify union members in safety and health instruction so they can then work within their own unions. The IUD provides trainees with *Facts & Analysis,* a newsletter devoted to job safety and health. Realizing the weakness of existing standards in the health area, the IUD has set up a program to develop information and submit requests for emergency temporary standards to the Department of Labor. These standards would be effective immediately if the Secretary found them necessary, and would then lead to a rule-making proceeding. The AFL-CIO proposed a tightening of the exposure limit to asbestos. The Labor Department has accepted this proposal by granting an emergency temporary standard. (However, the final asbestos standard, promulgated by the Labor Department on June 6, 1972, gives industry four years to meet the tighter limit.) Seventeen other requests, dealing with such exposures as lead, noise, and mercury, are in the process of being filed as a method of avoiding NIOSH's cumbersome criteria-package system.

The main complaint from the unions is lack of money. The AFL-CIO's policy of not accepting government grants means that they have to spend their own funds for their training institutes and cannot take advantage of Labor Department contracts;[66] only five AFL-CIO people are working part-time on safety and health in Washington. Yet at the last AFL-CIO Convention President George Meany received a salary hike to ninety thousand dollars, and the salary of the union's secretary-treasurer was increased to sixty thousand dollars, raises of 28 percent and 33 percent respectively[67] (pending review by the Pay Board set up under the wage-price freeze).

The Labor Department has expressed surprise that it has received only a little more than five hundred complaints from unions demanding inspections. Lawyers in the Department find very few unions contributing to the drafting of regulations or other regulatory procedures. They constantly hear from companies and state agencies, and thus are getting only one side of the story. Nor do the unions put any pressure on the Labor Department area or regional offices.

The costs of OSHA

As OSHA begins to affect conditions in the factories and work-shops, corporate officials are lamenting the costs that the act is imposing upon industry. Articles in business journals provide one medium for these pessimistic complaints. As one of them pointed out:

> Results of a survey conducted by *Safety Management* maga-zine indicate that the cost could run as high as three hundred dollars per employee for some plants.
>
> Cost per plant could be as high as eight-nine thousand dollars; and the overall cost to American industry will run into the billions of dollars.[68]

The January, 1972, issue of *Dun's* featured a sampling of the kind of costs industry perceives coming.[69] In the metal stamping industry, surveys on noise abatement have already cost $6 million. Producing items from car bodies to bottle tops, these companies are expecting to pay out over $6 billion in the next five years for compliance with federal standards.

The Home Builders Association estimates that OSHA will add between two thousand and three thousand dollars to the cost of every new home, while the paper and pulp industry reckons the tab at two hundred dollars per worker, or $140 million, over the next year or two.[70]

Some of these same articles, however, reveal a more opti-mistic picture, at least in the long run. Job injuries (not including industrial health disabilities) impair productivity to the tune of $9 billion a year.[71] James J. Sheehy, assistant treasurer at St. Regis Paper Co., has stated, "We are looking for at least a 14% increase in productivity [as a result of OSHA]. For a billion dollar company, this could add up to 3.5% to 4% more profit."[72] Consolidated Edison's safety director estimates, "OSHA will en-courage faster replacement of people by equipment, and of course safe plants will have lower accident rates and more productive manpower. The long-term impact will be an increase in produc-tivity industry-wide of at least 15%."[73]

Another long-term factor is the probable increase in the costs of workmen's compensation. It is very possible that political pressures will build for some kind of federal involvement in workmen's compensation. This, along with rising medical costs, could increase benefit levels and coverage, which would in turn drive up workmen's compensation insurance premiums.

These long-range prospects do not diminish an immediate problem: the threat that federal standards, if enforced, may cause some plants to shut down. The circumstances surrounding a recent closing suggest the need for a careful assessment of this aspect of the economic impact of OSHA. A plant manufacturing asbestos pipe-insulation products in Tyler, Texas, closed down, ostensibly because the Labor Department had fined the company (Pittsburgh Corning) $6,990 for exposing employees to "grossly excessive" levels of asbestos dust and had ordered the installation of new equipment that would reduce dust levels.[74] As a result, sixty-two union members lost their jobs. According to the OCAW, in the eighteen years the plant had been operating, about fifteen hundred workers had been exposed to the asbestos dust.[75] Pittsburgh Corning bought the plant in 1962, eight years after the previous owner had moved it to Texas from Paterson, New Jersey. Before the New Jersey operation terminated, Dr. Irving J. Selikoff had studied some of the workers and had found that men who had worked as little as one year in the plant had died of cancer at five times the expected U. S. death rate for cancer.[76]

Pittsburgh Corning claims that federal standards forced the plant to close. But the corporation has not revealed any economic data to support this allegation. OCAW officials suspect that far from causing the closing, health and safety standards were merely an excuse used by management to shut down a marginal operation because of low profits. (Indeed, in similar cases where companies claimed that plant closings were triggered by federal pollution standards, *New York Times* reporter Gladwin Hill found that the vast majority were closed for other reasons, with pollution standards serving as a convenient excuse.[77]) Moreover, any public interest evaluation of the costs of closing the Tyler plant would have to take other considerations into account. The men who lost their jobs in Tyler may have been spared the possibility of developing asbestosis and lung cancer, disabling illnesses that eventually would have forced them to seek some kind of income maintenance and medical benefits from workmen's compensation, welfare, or other public sources. These savings must be included on the balance sheet, and Congress might well consider converting the savings into some kind of supplemental unemployment insurance, so that workers would not be forced to bear the entire cost of genuine OSHA shutdowns.

An aspect of the total economic picture that may serve to offset the closing down of hazardous operations is the growth effect caused by the demand for safety and health equipment. Companies producing and marketing this equipment will expand and create new employment. For example, *Power* magazine recently reported, "Manufacturers are offering a wider selection of reduced-noise-making equipment. Low-noise fans, pumps, compressors, valves, gear boxes have all either been available or are recently developed items to help plants meet new regulations."[78]

The Occupational Safety and Health Administration's Office of Standards is currently collecting data in an attempt to assess the cost of the new standards. Although OSHA does not include cost as a factor for consideration in setting standards, Section 6(b)(5) does specify that the feasibility of standards must be considered, and feasibility arguably includes cost. Mr. Paul Preusser, the one-man economics department within the Office of Standards, is working on cost estimates for the asbestos standard. By contracting out a limited survey to experts in various asbestos companies, he hopes to obtain some picture of the possible costs and benefits of the standard.

Studies like this could provide important and useful data on the economic consequences of OSHA. However, the federal government has gone much further in attempting to graft economic considerations into the act. The Office of Management and Budget (OMB), an agency of the executive branch serving as fiscal watchdog over the activities of the federal government, has taken upon itself to evaluate proposed health and safety standards *before* they are published for comment in the *Federal Register*. According to a memorandum issued by OMB to all heads of federal departments and agencies on October 5, 1971, job safety and health (as well as other) regulations "which could be expected to . . . impose significant costs on, or negative benefits to, non-federal sectors" (a euphemism for private industry) must be submitted to OMB at least thirty days before publication in the *Federal Register*.

OSHA contains no provision allowing the pre-screening of proposed regulations by OMB or any other agency. It is crucial that OMB be held strictly accountable for any tinkering it may do with standards for which the Department of Labor has statutory responsibility. Any changes made by OMB, along with detailed supporting reasons, should be made public in the *Federal Register*—to facilitate an evaluation of the possibility that concern for corporate profits might be taking precedence over OSHA's mission to reduce the human and social costs of work accidents and diseases.

Counterattack on OSHA

The first serious attempt to cripple OSHA legislatively emerged from objections to the Labor Department's jurisdiction over businesses never before subject to effective regulation of any kind. On February 29, 1972, Senator Carl T. Curtis, a conservative Republican from Nebraska, introduced S. 3262, a series of amendments to the act. A similar bill, H.R. 13941, was offered in the House, and concurrently a barrage of complaints about the administration of OSHA, as it affected small establishments, began to appear in the *Congressional Record*.[79] These bills, which came to be known as the Curtis amendments, were ostensibly designed to meet the objections of small businessmen and farmers; but they also sought to gut OSHA's entire standard-setting process and impose cumbersome procedural burdens upon the Department of Labor.[80]

Amendments are normally referred to the same Congressional committees that produced the original bill. But in this instance, both the Senate Subcommittee on Labor and the House Select Subcommittee on Labor, along with organized labor, chose to ignore the pressures that were building, especially from the rural areas of the Midwest and West. "Don't worry," a pro-OSHA, pro-union Congressional staffer assured a member of the Study Group. "We control the labor committees, and we won't even give them a hearing."

This strategy of inaction backfired badly. A subcommittee of the House Select Committee on Small Business held hearings on the impact of OSHA on small businessmen. And at the same time, in a surprise move on the floor of the House, critics of OSHA secured the passage of an amendment to the 1973 Labor-HEW Appropriations Bill prohibiting the use of federal funds to pay federal inspectors to inspect businesses employing fewer than twenty-five workers. This effectively removed 90 percent of all previously covered employers and 30 percent of previously covered employees from federal inspection. A similar amendment limited to businesses with fewer than fifteen employees passed the Senate during its appropriations debate.

The Senate-House conference on the Appropriations Bill adopted the Senate version of the exemption. After President Nixon vetoed the entire bill for reasons unrelated to job health and safety, the House and Senate once again considered the small business exemption. This time both the labor unions and the Department of Labor strenuously opposed it, and oversight hearings by both Senate and House Labor Subcommittees focused on it. But the momentum produced by the protests of small businessmen proved too strong. The House passed an amendment exempting businesses with fewer than fifteen employees, and the Senate (after a heated debate during which Republican Senators Jacob Javits and Morris Cotton almost came to blows) lowered its exemption to businesses with three or fewer employees. The conference once again adopted the Senate version, but once again, President Nixon vetoed the bill.

Testimony before the subcommittee hearings revealed a combination of reasons for the uproar over OSHA: apprehension over potential standards enforcement on the part of employers never before subject to health and safety laws; plain unwillingness to submit to federal regulation; poor communication from the Labor Department's Occupational Safety and Health Administration; allegedly arbitrary action by some federal inspectors; and several allegedly unfair federal standards. Mail from constituents to Congressmen complained of "gestapo tactics" by federal inspectors, closing of businesses because of the high costs of compliance, and the levying of substantial penalties for violations of standards about which the companies fined knew nothing.

But small firms can be every bit as hazardous as large ones. As Jacob Clayman of the IUD pointed out in testimony before the House subcommittee:

An occupational health survey of the Chicago metropolitan area completed by the Department of Health, Education and Welfare in 1970 demonstrates what can be found in
small plants: the highest concentration of health hazards; the
fewest safeguards; and the least awareness, on the part of management, of the threat to life that exists in these workplaces.

The problem of physical hazards is equally horrifying.
Greater than 96 percent of the sixteen thousand logging
camps in this country have twenty or fewer employees.
Their disabling injury rate in 1970 was about three times the
rate among all manufacturing employers: nine injured employees for each one hundred. In Wyoming [where many of
the complaints against OSHA were originating], according to
the Bureau of Labor Statistics, it was twenty-four injured
workers per hundred in this industry.[81]

The information policies of the Occupational Safety and
Health Administration deserve criticism. No one can readily understand its administrative regulations and three hundred pages
of standards, with their numerous incorporated references. Requests for information can take up to six weeks, and replies often
direct the inquirer back to the *Federal Register*. Senator Curtis
received such a response when he requested information on
behalf of a constituent.[82] Unions are developing standards information for their members, as well as pamphlets and books
about employee rights and procedures for implementation. Small
business needs the same kind of service. Indeed, the AFL-CIO
has asked for the appropriation of additional funds under OSHA
for this purpose, but the Labor Department has refused to support this request.

The Report of the Commission

On July 31, 1972, the National Commission on State Workmen's Compensation Laws released its final report, and American
workers found themselves shortchanged once again. The Commission documented all the familiar inadequacies of the state acts,
but then proceeded to recommend against immediate federal action,
urging with sadly misplaced charity that the states be given three
more years to improve their laws. Senator Javits, whose efforts had
created the Commission, quickly disassociated himself from its

no-action recommendation, and has introduced a bill setting minimum federal standards for state compensation laws.

It is painfully clear that OSHA has not ushered in a Golden Era of job health and safety. Though the basic structure of the law furnishes a legal arsenal for a fresh assault on the scourge of industrial accidents and diseases, the Labor Department hesitates to mount a sorely needed offensive, but rather tends to see the act in terms of loopholes supporting a policy of giving relief to employers. The rapid surrender of authority to the states is evidence of the Department's reluctance to take the initiative. At the same time, organized labor has not yet seized the opportunities presented by OSHA to mobilize the support and participation of the rank and file and the general public.

In its historical perspective, the present plight of the movement for job health and safety smacks of the same old story. It still has not realized its enormous potential for justice in the occupational environment. But the possibilities for achieving this goal may be brighter now because of the enactment of OSHA and, more important, because of the growing awareness on the part of American workers that they are continuing to pay an unnecessary, unreasonable, unconscionable price for the nation's industrial production and growth.

10

Conclusions and Recommendations

On December 11, 1971, almost one year after OSHA was passed, a natural gas explosion in a water tunnel dug beneath Lake Huron took the lives of twenty-two construction workers and injured nine others.[1] Several days after the tragedy, a former safety inspector for the Detroit Metropolitan Water System, which was building the tunnel, accused his superiors of allowing safety violations to go unchecked so that they could adhere to their time schedule.[2] Newspaper photos and accounts of the disaster once again reminded the public of the gravity, persistence, and ubiquity of job safety and health hazards.*

The history of the movement to protect workers from industrial accidents and diseases consists of a pattern of neglect interrupted by occasional efforts on the part of individuals and groups to come to grips with the problem. This book has described the degree to which these individuals and groups have fallen short of their goal.

OSHA, despite its imperfections and complexities, provides

---

* On December 23, 1971, the Occupational Safety and Health Administration obtained from a federal district court a temporary restraining order issued under OSHA to halt work at the tunnel site because of imminent danger. This was the first time an imminent hazard had been enjoined under Section 13 of the Act.

a new legal mechanism that aggressive officials within both organized labor and government can utilize in a sustained assault on the hazards of the industrial environment. But a glimpse at what has actually been done over the past twelve months indicates that the opportunities created by OSHA have scarcely been tapped.

Throughout this book, the authors have made numerous criticisms and have suggested steps that should be taken in order to bring about a meaningful reduction in the human toll exacted by occupational accidents and diseases. In this concluding chapter, we shall present further specific recommendations.

Our study leads us to the conclusion that mere tinkering with traditional approaches to job safety and health will eventually lead to the same old frustrations. What is necessary is an entirely fresh focus by all sectors of the movement. In our view, the key to real progress lies in the cultivation of worker involvement.

The employee's own life and health are at stake in the workplace. Truly adequate protection will not develop until the individual worker becomes aware of all the dangers of the workplace, their dimensions, and the alternatives that can furnish reasonable safeguards. The challenge, therefore, is to make occupational safety and health a vital, central part of the worker's everyday life, rather than an outside force generated by others, sweeping periodically through the plant, and then disappearing. The usual approaches to the problem have always smacked of paternalism, with workers being admonished and taken care of by company, government, and union officials. Our conviction is that the worker must be encouraged to take care of himself and participate in the control of the quality of his or her own occupational environment.

We advocate the development of worker citizenship, placing upon the individual worker maximum opportunity to share in the maintenance of a safe and healthful workplace. The boredom and discontent afflicting many of today's workers would substantially diminish in the face of a chance to engage in a fundamental form of industrial democracy.

Outsiders simply cannot and do not have the same interest as the worker in job safety and health. They do not have to spend their working lives amidst the fumes, gases, noise, and dust. Nor do they enjoy the same means of access that can be utilized to pressure corporate interests. But these limitations do not affect the struggle against in-plant pollution. The labor movement possesses a wide spectrum of action plans with which to approach job safety and health issues. Collective bargaining, grievance procedures, and daily contact with the actual hazard-spawning machinery and substances give workers a decided advantage that outsiders lack.

In a counterattack against the conservation movement, apologists for industry have sought to drive a wedge between workers and environmentalists by stressing that the imposition of anti-pollution measures by the environmentalists will result in a loss of jobs. This same tactic will have less chance of succeeding against workers seeking protection from job hazards if companies are forced to disclose all their safety-and-health-cost data in net terms, and the workers have access to professional aid in calculating for themselves the full cost of injuries and diseases to employees.

The men and women in the factories and workshops need outside help in the fight against industrial hazards. The efforts of government and union officials, professionals, students, the general public, and enlightened corporate officials are essential ingredients in the movement for job safety and health. But their efforts should be directed toward the fundamental goal of worker education and worker control of the industrial environment. Until employees accept the responsibility for initiating solutions to their own occupational health and safety problems, job hazards will continue to overwhelm laws, regulations, and voluntary measures to safeguard the work force.

It is also well to keep in mind that private individuals and groups committed to public-interest activity cannot externally impose their solutions upon the workers. Rather, an across-the-table mutual learning process should generate the energy for change. Any hint of an elitist attitude on the part of students and intellectuals, such as an insensitivity to the power of management over workers encouraged to challenge harmful practices, will provoke grass-roots resistance and cannot mobilize the necessary enthusiasm and commitment among rank-and-file workers.

With this in mind, let us turn to some specific recommendations.

NIOSH

It is essential that the National Institute of Occupational Safety and Health (NIOSH) not fall into the same pattern of inaction and ineffectiveness as its predecessor, the Bureau of Occupational Safety and Health (BOSH). NIOSH officials and technicians need to attract attention and support from the upper levels of the HEW bureaucracy, the general public, and the workers whom the institute was designed to serve.

One obvious starting point is the 1965 Frye Report, "Protecting the Health of Eighty Million Americans: A National Goal for Occupational Health."[3] NIOSH should dust off the report, update it (keeping in mind that it was admittedly inadequate in its own day), and then push aggressively for the implementation of its suggestions, in terms of personnel and funding. Top HEW officials may be aware of these priority needs, but they need to feel pressure to do something, so that they can in turn give the occupational health program credibility with Congress.

NIOSH is also in a position to publicize the 1970 Environmental Education Act, which has an authorized $23 million appropriation for fiscal year 1973, and is administered by HEW's Office of Education. Funds available under this act could be used to train labor personnel in occupational health, as well as to finance library resources in cities and schools. It is just as important for children to develop an awareness of occupational environment hazards as it is for them to learn about air and water pollution.

NIOSH itself has to bridge the gap between scientific professionalism and the workplace. Additional funds and more scientists will not help unless people at NIOSH are able to communicate with workers. To this end, NIOSH must cultivate in the minds of employees the image of a willing listener and a source of help. Curt letters dismissing inquiries because they were sent to the wrong office illustrate perfectly the worst way to deal with workers. NIOSH could gain much insight by adopting the OCAW strategy of talking with the workers in particular plants to learn what they perceive as health and safety hazards.

It is not a valid excuse to say that workers or unions are

not interested, particularly when they are kept unaware of the facts and risks to health. NIOSH officials have a responsibility to attract their interest. The Scientists Committee on Occupational Health, whose activities have been described in Chapter 9, provides one example of how medical doctors, chemists, and other scientists are making it their business to communicate with workers. A beginning would be for NIOSH to create a task force to develop approaches to workers. The institute is already loaning equipment to the Scientists Committee. Further contacts between NIOSH and such groups of scientists interested in occupational health problems would be advantageous to all involved.

Another modest step would be for NIOSH to draw up in comprehensible language a list of NIOSH offices and the responsibilities of each office. Employees and employers would then know whom to contact for toxicity requests and information on standards development and inspections. In order to assure wide circulation of this information, NIOSH could publish notices of the availability of the list in local union newspapers, which reach the rank and file on a regular basis and can be excellent media for communication. If some union newspapers are reluctant to print this information, NIOSH could publish it in leaflet form for public distribution.

The institute must give serious consideration to the rapid development of criteria for standards. The Secretary of HEW, through NIOSH, can issue regulations under Section 20(a)(5) of OSHA establishing a monitoring system for potentially toxic substances. There is also nothing to prevent the institute from publishing in the *Federal Register* preliminary drafts of criteria packages for standards—subjecting them to public comment so that experts outside NIOSH could participate in the development of these standards.

Finally, NIOSH should view itself as more than a research organization, and assume more responsibility for urging enforcement of health standards. One exemplary instance of such an assumption of responsibility has been recorded in the case of a Texas plant exposing workers to asbestos dust. The closing of the plant is described in chapter 9. Labor Department inspectors went into the plant at the urging of the OCAW. The reason the OCAW was able to present a strong case against the health hazards in the plant was that several NIOSH technicians had become aware of the hazards, and of the fact that the Labor Department had information about the hazards and was not going to initiate an inspection, and feeling outraged at this callous inaction, they passed information about the hazards to the OCAW.[4]

Department of Labor

This book has already directed specific criticisms at the Department of Labor for its pro-industry, antilabor efforts during the battle for OSHA and in the first year of its operation. The fundamental problem with the Department stems from its policies and attitudes toward job safety and health.

Thus, the Department has consistently stretched its interpretations of OSHA to the furthest limits of its language and legislative history in order to avoid imposing strict, immediate requirements upon employers and the states. Sweeping advance-notice regulations and the Department's refusal to recognize the importance of walk-around inspections are two examples cited in Chapter 9. In addition, the Department has followed the practice of retaining, in regulations promulgated under OSHA, as much discretion as possible in the Secretary of Labor. An instance of the tenacity with which the Department seeks to maximize its capacity to exercise discretion may be found in the regulations dealing with fines for the violation of posting requirements. A subsection of the penalty section of the act specifically sets a mandatory fine of up to one thousand dollars for an employer's failure to post whatever information is required to be posted by the act and regulations issued thereunder. But in all its regulations on posting, the Department has steadfastly refused to cite this specific subsection, thus intimating that it considers fines for posting violations discretionary. Posting is now the main source of employee information about inspection results and employee rights under the act. There is no excuse for not posting this information and no reason why the penalty for failing to post should not be mandatory.

Assigning the main training contract under OSHA to Boeing, and setting up an industry information officer but not a union information officer, lend credence to organized labor's insistence that the Labor Department is a hostile force. It may well be that nothing can be done about this under a Republican administration. Nonetheless, we would like to suggest the following specific recommendations:

1. It is clear that the Labor Department's Occupational Safety and Health Administration needs more inspectors. By June, 1972, 500 inspectors were scheduled to be hired, an increase of 187 over the inspection force on the job in October, 1971. This means that at the current rate (an estimated seventy thousand inspections a year), it will take fifty-eight years to reach the 4.1 million establishments covered by OSHA. One way to expand the inspection force is to modify one of the prerequisites for the job. The Labor Department has been requiring that inspectors meet civil service qualifications that eliminate many men and women who have gained practical experience within the trade union movement.

2. The Labor Department has authority to issue self-inspection regulations for employers, requiring that reports be filed with the Department. The issuance of such regulations would give the Department some idea of conditions in plants that cannot be officially visited because of the shortage of inspectors.

3. Current regulations on advance notice allow employers to notify employees about impending inspections and thus reduce contact between federal inspectors and employee representatives. The inspectors themselves should make these notifications, so that they can become as familiar with employee representatives as they are with company officials.

4. The Labor Department poster notifying employees of their rights cries out for redesigning. It is extremely unattractive and noncommunicative. Labor educators might have some better ideas for the Labor Department concerning methods of imparting this and other information to workers.

5. The Department should make an effort to include workers and independent critics in advisory positions and on committees.

6. Access to information within the Labor Department is still a problem. One positive development has been the Department's publication of its Compliance Manual through the U.S. Government Printing Office. But no decision has yet been made on the public availability of OSHA inspection reports and other

forms required to be filled out by inspectors. It would be a waste of time and effort for the Department to provoke another Freedom of Information Act lawsuit for the "liberation" of this information.

7. The current practice is for inspectors to confer with company officials after an inspection. Since employees are supposed to participate in inspections to the same degree as employers, and may even have requested the particular visit, there is every reason why employee representatives should be permitted to participate in these conferences. Such discussions may eliminate future disputes over citations and abatement periods.

8. The Labor Department should take a restrictive view of requests for lenience and variances by employers. The Department's responsibility is not to protect employers from standards but to protect employees from unsafe and unhealthful working conditions. If the Department built up worker support instead of alienating it, it would be in a stronger position to counter corporate pressure to ease up in the enforcement of the act.

9. The Labor Department has authority under the act to educate, train, and consult employees as well as employers. Its current practice is to conduct one-day sessions on the law and regulations for both union and company officials. A more aggressive approach would be to hire organizers and professionals to go around and sell OSHA to the workers. As suggested previously for NIOSH, the Labor Department could take advantage of the union press to spread information on safety and health developments. While the argument will be made that the Labor Department should not involve itself in labor disputes over health and safety matters, this does not preclude making information available to both sides. The Department is surely sensitive enough to avoid direct interference in strike situations, without the need for withdrawing from all involvement with the workers.

10. Federal inspectors might benefit from instruction by union organizers in the regions where they are assigned, so that they can become familiar with some of the problems the unions face. This instruction could be included within the Labor Department's training program. Another way to enrich the program would be to encourage veteran inspectors to suggest changes reflecting their own practical experience.

11. How state plans propose to involve individual workers in job safety and health activity merits careful scrutiny by the Labor Department. Worker-education materials submitted by the states can be evaluated as evidence of the effectiveness of a state plan.

12. The Labor Department should take a strong stand in favor of the protection of employee rights in nonunion plants. The training of inspectors ought to cover dealing with employees who enjoy no union representation. An essential safeguard for these employees is to assure them the opportunity to communicate with and complain to federal inspectors under circumstances that minimize the possibility of employer retaliation.

13. The Department should re-evaluate its enforcement priorities to give greater emphasis to the more serious of the health hazards, such as asbestosis and byssinosis, as described in Chapter 2 of this book. Better coordination with NIOSH might be one way to achieve a factual basis for such a shift in priorities.

The unions

Rank-and-file participation in the control of the industrial environment requires solid support from the health and safety bureaucracies in the international and local unions. The task at hand is to develop structures that will foster worker responsibility and will enable individual employees to protect themselves according to their self-perceived needs.

The worker in the plant is well placed to promote the implementation of OSHA and related safety and health efforts. The challenge is to train him and to provide him with powerful outside political, professional, and educational support. Thus equipped and fortified, the individual worker can check on compliance dates, question injury statistics, and watch the posting of information by the company. He can help gather information essential to collective bargaining. He can learn what kind of medical, legal, and scientific services he lacks and how to obtain them locally. The development of community support can best be achieved by grass-roots organizing led by workers who live in the community involved.

The international unions have an essential role to play.

They should stop complaining about the Labor Department's shortage of inspectors long enough to do something about the small size of their own safety and health staffs. The UAW, with 1.4 million members, and the Steelworkers, with 1.25 million members, have only four safety and health men each, while the OCAW, the Industrial Union Department, and the AFL-CIO have only one. The unions have Washington lobbyists who work effectively in Congress, the Labor Department, and HEW. At least one of these lobbyists in each union could work full time on job safety and health. The legal complexities of OSHA have created a pressing need for the formation of a group of lawyers, aided by physicians, engineers, and scientists, to represent the interests of the labor movement in overseeing the implementation of the act. By late 1972, the Labor Department was employing more than twenty attorneys to do full-time legal work for the Occupational Safety and Health Administration. At the same time there was only one lawyer in Washington, affiliated with Ralph Nader's Health Research Group, working full-time to oversee the output of all these attorneys. The international unions continue to ignore the vital necessity of utilizing legal talent to advance the interests of workers at points within the administrative decision-making process.

The unions could also make more effective use of their regional, district, and local offices. For the larger unions, each region should have a full-time safety and health person to keep watch over the Labor Department regional offices that administer OSHA. Labor Department area offices can be handled by local unions or regional offices, or both, depending upon the size of the local and of the region. A similar type of organizational structure could keep state safety and health programs under surveillance— programs that will be of critical importance in states whose job safety and health plans have been approved by the Labor Department. Various unions can combine resources within an area to develop educational programs and training in hazard recognition, as well as for the presentation of testimony before state agencies and the fashioning of strategies for dealing with state officials.

An approach combining grass-roots organizing with the efforts of a safety and health bureaucracy requires the fashioning of educational programs for the rank and file. Staff people must familiarize themselves with both federal and state laws and regulations. AFL-CIO training institutes at universities and in regional offices provide a model for these courses. The employee representatives designated by the unions for walk-around inspections and for dealing with the Labor Department will need extensive training not only in the law but also in some of the specific medical, scientific, economic, and legal problems that relate to their own workplaces. Union educators could help by developing training games that call for the use of OSHA to solve problems that may arise in the plants.

A challenging task for local and regional union offices is to find the technical resources to support safety and health activities. The Medical Committee on Human Rights and the Scientists Committee on Occupational Health are examples of support groups that could be approached. Universities may also provide resources, but it is important that their support be enlisted on union terms to fulfill what the unions perceive as their own needs.

Because dues stretch only so far, the unions will have to seek funds for training and research outside of the labor movement. Foundation grants and government contracts are possible sources. For example, the National Commission on Productivity, set up under the recent wage-price-controls system, has at its disposal until April 30, 1973, $10 million to assist labor and management in programs designed to reduce waste and absenteeism and to promote worker safety and health. The AFL-CIO could compile a list of these sources of funds.

Use of the mass media can be of critical importance in the struggle for safe and healthful workplaces. Hence part of the training afforded regional and local officials should include the theory and practice of mass communications. The labor press can help keep attention focused on the occupational environment by encouraging exchanges of information and success stories. Since impediments to the free flow of information may develop within union bureaucracies, the establishment of a regional or national hot-line would enable rank-and-file members to bypass any inactive or recalcitrant locals and give them direct access to vital facts and figures.

Union tactics do not have to be bound by OSHA. Traditional methods of collective bargaining and organizing merit full utilization. Collective bargaining can extend the protections of OSHA into the daily life of a plant. Unions should seek the right to strike over safety and health without fear of discharge and with back pay if conditions turn out to be truly hazardous. Joint safety and health committees have the potential of maintaining constant pressure on companies to fulfill requirements set by OSHA. The unions should also challenge corporate secrecy whenever companies raise the issue of the costs of safety and health protection but refuse to reveal specific data. Another way to deal with the cost issue is to seek the help of economists who could analyze all available data and compare the costs of preventing accidents and diseases with their human costs. Only 20 million workers are unionized, yet the protections of OSHA are geared to the participation of organized labor. Therefore, the unions have good reason to stress the job safety and health issue in the course of organizing drives.

Private groups

A recent article in the magazine *Occupational Hazards* reported that although OSHA has greatly increased the need for professional safety and health engineers within industry, students are not entering the field because of their skepticism about corporate commitment to the welfare of employees.[5] This attitude on the part of idealistic young people is the fruit of the traditional subservience to corporate interests displayed by company doctors, scientists, and engineers, and their professional associations. The promotion of corporate profits has inspired the great majority of these in-house professionals to neglect their higher responsibility to the public good, and hence it is not surprising that young men and women are hardly attracted to this perversion of health and safety work.

An obvious escape from this sorry state of affairs is to sever the financial dependence of these professionals upon industry, so that they can exercise truly professional judgment. For example, doctors and nurses working in factories could be hired jointly by management and the union, or they could be completely independent, in the nature of public health officers. Private research groups, instead of being financed by industry, could be funded through a joint labor-management-public commission.

Universities have a responsibility to participate in the movement for occupational safety and health. More medical schools should teach occupational health and stress preventive techniques. Students could do clinical work in factories. Law students, especially in state capitals, have the opportunity to probe and evaluate the administration and enforcement of job safety and health laws and regulations and, in conjunction with medical and other health science students, engage in legal action projects wherever necessary and feasible. It is a sad commentary that in the preparation of this book, the Study Group uncovered only one academic in the whole country, Dr. George Hagglund of the University of Wisconsin's School for Workers, who was devoting himself to a constant, critical evaluation of how his state's industrial safety and health laws are administered.

Finally, student-financed Public Interest Research Groups (PIRG's) at the state level can furnish substantial support to the job health and safety movement. A suggested framework for such a PIRG project might include:

1. The development of on-going, self-sufficient safety and health programs within the existing union structure in several plants, with emphasis on worker independence. The project could:

    (a) Establish an accident and health-hazard reporting system, through the shop stewards, to educate and involve workers. An International Union of Electrical, Radio and Machine Workers local in Lynn, Massachusetts, is using such a program. This pilot experiment resulted in over forty accident reports, seventeen of which led to hospital attention, and in the recording of forty-five health-hazard reports.

    (b) Put together a report on plant hazards by taking a hazard survey, along with a shop poll, on which hazards workers think are important. Again, shop stewards and the rank and file can gather data and begin basic hazard analysis. Workers can use this record to support requests for safety and health inspections, grievance and workmen's compensation proceedings, and collective bargaining demands. A Teamsters local in St. Louis has begun this kind of program.

(c) Publicize all this information in the union newspaper. The international union can then use the data in these articles in dealing with the Labor Department in Washington.

(d) Run training sessions on the laws, regulations, and technical data. OSHA provides funds for training courses and could be tapped as a source of financing for these sessions.

2. Organization of safety and health activities in nonunion plants. The project could:

(a) Set up a safety and health committee.

(b) Have people on this committee serve as employee representatives, to deal with inspectors and make requests for information.

(c) Implement a reporting system that would provide protection for workers against employer retaliation.

3. The creation of structures to bring together workers, their families, community groups, and the universities in a coalition to protect the workers' interest in controlling the quality of their lives. The project could:

(a) Offer to the community movies, newspaper articles, and lectures on the hazards found both inside and outside the workshop.

(b) Build up local ecology centers and library resources and publicize the availability of existing materials.

(c) Form a community-wide committee to deal with environmental issues that affect both the worker and the community as a whole.

4. The monitoring of state safety and health programs, to make sure that the state has the best plan possible and that it is enforced. The project could work with state agencies to develop new standards for safety and health and to support increases in budget and personnel.

Congress

The responsibility of the Congress is to assure that the job safety and health activities of the federal government receive adequate funding, and to oversee the administration and enforcement of OSHA with a view toward considering any amendments that appear necessary to strengthen the act. We have already discussed the need for increased appropriations. At this point, we shall suggest several amendments that merit scrutiny by Congress.

1. We have noted the Labor Department's debilitating interpretation of the phrase "at least as effective" in Section 18(c) of the act. A change in language could make it unmistakably clear that federal standards and rights are to be considered *minimum* levels for state plans, which must contain at least those standards and rights specified in OSHA and in any regulations promulgated thereunder.

2. Labor Department compliance officers should have red-tag authority under Section 13, enabling them to shut down a machine or terminate a process immediately if they find an imminent danger of death or serious physical harm, subject to review by the courts. The original bill proposed by the Johnson administration in 1968 granted this authority.[6] An alternative approach would be to vest this power in the Secretary of Labor. The Secretary of HEW has authority to seize household products and toys deemed to present an imminent hazard to public health. The American worker deserves just as much protection in emergency circumstances.

3. Section 9(b) requires that employers post prominently in the workplace a citation of violations issued by the Secretary of Labor. There is no requirement that the citation be posted immediately—a loophole that ought to be plugged.

4. Section 20(a)(5) gives the Secretary of HEW discretion to issue regulations on the monitoring of potentially toxic substances. Congress should scrutinize the performance of HEW and, if necessary, amend the act to require HEW to promulgate monitoring regulations within a definite period of time.

5. If a factory or workplace is forced to suspend all or part of its operation because of any enforcement proceeding under OSHA, the employees ought not to suffer any economic hardship.

Therefore, Congress should consider an amendment giving employees affected by safety and health shutdowns the right to be transferred to another job at the same (or higher) rate of pay or, if no jobs are available, the right to be paid in full during the time it takes to eliminate the hazards.

The goal of the job health and safety movement should be to put scientific knowledge and modern technology to the humane task of eliminating from the workplace all unreasonable and unnecessary hazards to life and limb. This is not going to happen overnight. It will take sustained effort over a considerable period of time and a total mobilization of available resources to counter the encrustations of decades of neglect and the ever-burgeoning perils spawned by modern industry.

History teaches the folly of overreliance on bureaucracy. Whether in federal or state agencies or in labor unions, bureaucratic machinery cannot carry the primary burden in the long-term fight against occupational accidents and diseases. The motivating force must come from the workers themselves. Armed with an understanding of the daily hazards they face and the legal tools available to them under OSHA, workers can struggle to share in the exercise of significant control over the quality of their own lives.

The American labor movement has fought long and arduous battles in winning fundamental rights for working men and women. Job health and safety is perhaps the most fundamental of rights, involving both life itself and such basic concomitants as industrial democracy and the elevation of the dignity of work.

The life of Albert Hutchinson should serve as an inspiration and a reminder for both union leaders and the rank and file. An asbestos worker from the age of seventeen, he rose through the ranks of his union to become president of the International Association of Asbestos Workers. Learning of the tragic effects of the inhalation of asbestos dust, he placed himself in the forefront of efforts to reduce the hazard to which he had been exposed for many years. He was an active participant in the unsuccessful attempt to persuade the Labor Department to move quickly to tighten the asbestos exposure standard. On June 19, 1972—twelve days after the Occupational Safety and Health Administration rejected the recommendations of its own advisory committee and scientists from NIOSH and gave industry more than four years to reduce in half the deadly dust inhaled by asbestos workers—Albert Hutchinson, at the age of sixty-one, succumbed to the ravages of lung cancer, the very disease from which he was striving to spare the members of his union.

Appendix

The Occupational Safety and Health Act of 1970

AN ACT

To assure safe and healthful working conditions for working men and women; by authorizing enforcement of the standards developed under the Act; by assisting and encouraging the States in their efforts to assure safe and healthful working conditions; by providing for research, information, education, and training in the field of occupational safety and health; and for other purposes.

Be it enacted by the Senate and House of Representatives of the United States of America in Congress assembled, That this Act may be cited as the "Occupational Safety and Health Act of 1970".

CONGRESSIONAL FINDINGS AND PURPOSE

SEC. (2) The Congress finds that personal injuries and illnesses arising out of work situations impose a substantial burden upon, and are a hindrance to, interstate commerce in terms of lost production, wage loss, medical expenses, and disability compensation payments.

(b) The Congress declares it to be its purpose and policy, through the exercise of its powers to regulate commerce among the several States and with foreign nations and to provide for the general welfare, to assure so far as possible every working man and woman in the Nation safe and healthful working conditions and to preserve our human resources—

(1) by encouraging employers and employees in their efforts to reduce the number of occupational safety and health hazards at their places of employment, and to stimulate employers and employees to institute new and to perfect existing programs for providing safe and healthful working conditions;

(2) by providing that employers and employees have separate but dependent responsibilities and rights with respect to achieving safe and healthful working conditions;

(3) by authorizing the Secretary of Labor to set mandatory occupational safety and health standards applicable to businesses affecting interstate commerce, and by creating an Occupational Safety and Health Review Commission for carrying out adjudicatory functions under the Act;

(4) by building upon advances already made through employer and employee initiative for providing safe and healthful working conditions;

(5) by providing for research in the field of occupational safety and health, including the psychological factors involved, and by developing

innovative methods, techniques, and approaches for dealing with occupational safety and health problems;

(6) by exploring ways to discover latent diseases, establishing causal connections between diseases and work in environmental conditions, and conducting other research relating to health problems, in recognition of the fact that occupational health standards present problems often different from those involved in occupational safety;

(7) by providing medical criteria which will assure insofar as practicable that no employee will suffer diminished health, functional capacity, or life expectancy as a result of his work experience;

(8) by providing for training programs to increase the number and competence of personnel engaged in the field of occupational safety and health;

(9) by providing for the development and promulgation of occupational safety and health standards;

(10) by providing an effective enforcement program which shall include a prohibition against giving advance notice of any inspection and sanctions for any individual violating this prohibition;

(11) by encouraging the States to assume the fullest responsibility for the administration and enforcement of their occupational safety and health laws by providing grants to the States to assist in identifying their needs and responsibilities in the area of occupational safety and health, to develop plans in accordance with the provisions of this Act, to improve the administration and enforcement of State occupational safety and health laws, and to conduct experimental and demonstration projects in connection therewith;

(12) by providing for appropriate reporting procedures with respect to occupational safety and health which procedures will help achieve the objectives of this Act and accurately describe the nature of the occupational safety and health problem;

(13) by encouraging joint labor-management efforts to reduce injuries and disease arising out of employment.

DEFINITIONS

SEC. 3. For the purposes of this Act—

(1) The term "Secretary" means the Secretary of Labor.

(2) The term "Commission" means the Occupational Safety and Health Review Commission established under this Act.

(3) The term "commerce" means trade, traffic, commerce, transportation, or communication among the several States, or between a State and any place outside thereof, or within the District of Columbia, or a possession of the United States (other than the Trust Territory of the Pacific Islands), or between points in the same State but through a point outside thereof.

(4) The term "person" means one or more individuals, partnerships, associations, corporations, business trusts, legal representatives, or any organized group of persons.

(5) The term "employer" means a person engaged in a business affecting commerce who has employees, but does not include the United States or any State or political subdivision of a State.

(6) The term "employee" means an employee of an employer who is employed in a business of his employer which affects commerce.

(7) The term "State" includes a State of the United States, the District of Columbia, Puerto Rico, the Virgin Islands, American Samoa, Guam, and the Trust Territory of the Pacific Islands.

(8) The term "occupational safety and health standard" means a standard which requires conditions, or the adoption or use of one or more practices, means, methods, operations, or processes, reasonably necessary or appropriate to provide safe or healthful employment and places of employment.

(9) The term "national consensus standard" means any occupational safety and health standard or modification thereof which (1), has been adopted and promulgated by a nationally recognized standards-producing organization under procedures whereby it can be determined by the Secretary that persons interested and affected by the scope or provisions of the standard have reached substantial agreement on its adoption, (2) was formulated in a manner which afforded an opportunity for diverse views to be considered and (3) has been designated as such a standard by the Secretary, after consultation with other appropriate Federal agencies.

(10) The term "established Federal standard" means any operative occupational safety and health standerd established by any agency of the United States and presently in effect, or contained in any Act of Congress in force on the date of enactment of this Act.

(11) The term "Committee" means the National Advisory Committee on Occupational Safety and Health established under this Act.

(12) The term "Director" means the Director of the National Institute for Occupational Safety and Health.

(13) The term "Institute" means the National Institute for Occupational Safety and Health established under this Act.

(14) The term "Workmen's Compensation Commission" means the National Commission on State Workmen's Compensation Laws established under this Act.

APPLICABILITY OF THIS ACT

SEC. 4. (a) This Act shall apply with respect to employment performed in a workplace in a State, the District of Columbia, the Commonwealth of Puerto Rico, the Virgin Islands, American Samoa, Guam, the Trust Territory of the Pacific Islands, Wake Island, Outer Continental Shelf lands defined in the Outer Continental Shelf Lands Act, Johnston Island, and the Canal Zone. The Secretary of the Interior shall, by regulation, provide for judicial enforcement of this Act by the courts established for areas in which there are no United States district courts having jurisdiction.

(b) (1) Nothing in this Act shall apply to working conditions of employees with respect to which other Federal agencies, and State agencies acting under section 274 of the Atomic Energy Act of 1954, as amended (42 U.S.C. 2021), exercise statutory authority to prescribe or enforce standards or regulations affecting occupational safety or health.

(2) The safety and health standards promulgated under the Act of June 30, 1936, commonly known as the Walsh-Healey Act (41 U.S.C. 35 et seq.), the Service Contract Act of 1965 (41 U.S.C. 351 et seq.), Public Law 91–54, Act of August 9, 1969 (40 U.S.C. 333), Public Law 85–742, Act of August 23, 1958 (33 U.S.C. 941), and the National Foundation on Arts and Humanities Act (20 U.S.C. 951 et seq.) are superseded on the effective

date of corresponding standards, promulgated under this Act, which are determined by the Secretary to be more effective. Standards issued under the laws listed in this paragraph and in effect on or after the effective date of this Act shall be deemed to be occupational safety and health standards issued under this Act, as well as under such other Acts.

(3) The Secretary shall, within three years after the effective date of this Act, report to the Congress his recommendations for legislation to avoid unnecessary duplication and to achieve coordination between this Act and other Federal laws.

(4) Nothing in this Act shall be construed to supersede or in any manner affect any workmen's compensation law or to enlarge or diminish or affect in any other manner the common law or statutory rights, duties, or liabilities of employers and employees under any law with respect to injuries, diseases, or death of employees arising out of, or in the course of, employment.

DUTIES

SEC. 5. (a) Each employer—

(1) shall furnish to each of his employees employment and a place of employment which are free from recognized hazards that are causing or are likely to cause death or serious physical harm to his employees;

(2) shall comply with occupational safety and health standards promulgated under this Act.

(b) Each employee shall comply with occupational safety and health standards and all rules, regulations, and orders issued pursuant to this Act which are applicable to his own actions and conduct.

OCCUPATIONAL SAFETY AND HEALTH STANDARDS

SEC. 6. (a) Without regard to chapter 5 of title 5, United States Code, or to the other subsections of this section, the Secretary shall, as soon as practicable during the period beginning with the effective date of this Act and ending two years after such date, by rule promulgate as an occupational safety or health standard any national consensus standard, and any established Federal standard, unless he determines that the promulgation of such a standard would not result in improved safety or health for specifically designated employees. In the event of conflict among any such standards, the Secretary shall promulgate the standard which assures the greatest protection of the safety or health of the affected employees.

(b) The Secretary may by rule promulgate, modify, or revoke any occupational safety or health standard in the following manner:

(1) Whenever the Secretary, upon the basis of information submitted to him in writing by an interested person, a representative of any organization of employers or employees, a nationally recognized standards-producing organization, the Secretary of Health, Education, and Welfare, the National Institute for Occupational Safety and Health, or a State or political subdivision, or on the basis of information developed by the Secretary or otherwise available to him, determines that a rule should be promulgated in order to serve the objectives of this Act, the Secretary may request the recommendations of an advisory committee appointed under section 7 of this Act. The Secretary shall provide such an advisory committee with any proposals of his own or of the Secretary of Health, Education, and Welfare, together with all pertinent factual information developed by the Secretary or the Sec-

retary of Health, Education, and Welfare, or otherwise available, including the results of research, demonstrations, and experiments. An advisory committee shall submit to the Secretary its recommendations regarding the rule to be promulgated within ninety days from the date of its appointment or within such longer or shorter period as may be prescribed by the Secretary, but in no event for a period which is longer than two hundred and seventy days.

(2) The Secretary shall publish a proposed rule promulgating, modifying, or revoking an occupational safety or health standard in the Federal Register and shall afford interested persons a period of thirty days after publication to submit written data or comments. Where an advisory committee is appointed and the Secretary determines that a rule should be issued, he shall publish the proposed rule within sixty days after the submission of the advisory committee's recommendations or the expiration of the period prescribed by the Secretary for such submission.

(3) On or before the last day of the period provided for the submission of written data or comments under paragraph (2), any interested person may file with the Secretary written objections to the proposed rule, stating the grounds therefor and requesting a public hearing on such objections. Within thirty days after the last day for filing such objections, the Secretary shall publish in the Federal Register a notice specifying the occupational safety or health standard to which objections have been filed and a hearing requested, and specifying a time and place for such hearing.

(4) Within sixty days after the expiration of the period provided for the submission of written data or comments under paragraph (2), or within sixty days after the completion of any hearing held under paragraph (3), the Secretary shall issue a rule promulgating, modifying, or revoking an occupational safety or health standard or make a determination that a rule should not be issued. Such a rule may contain a provision delaying its effective date for such period (not in excess of ninety days) as the Secretary determines may be necessary to insure that affected employers and employees will be informed of the existence of the standard and of its terms and that employers affected are given an opportunity to familiarize themselves and their employees with the existence of the requirements of the standard.

(5) The Secretary, in promulgating standards dealing with toxic materials or harmful physical agents under this subsection, shall set the standard which most adequately assures, to the extent feasible, on the basis of the best available evidence, that no employee will suffer material impairment of health or functional capacity even if such employee has regular exposure to the hazard dealt with by such standard for the period of his working life. Development of standards under this subsection shall be based upon research, demonstrations, experiments, and such other information as may be appropriate. In addition to the attainment of the highest degree of health and safety protection for the employee, other considerations shall be the latest available scientific data in the field, the feasibility of the standards, and experience gained under this and other health and safety laws. Whenever practicable, the standard promulgated shall be expressed in terms of objective criteria and of the performance desired.

(6) (A) Any employer may apply to the Secretary for a temporary order granting a variance from a standard or any provision thereof promulgated under this section. Such temporary order shall be granted only if the employer files an application which meets the requirements of clause (B) and

establishes that (i) he is unable to comply with a standard by its effective date because of unavailability of professional or technical personnel or of materials and equipment needed to come into compliance with the standard or because necessary construction or alteration of facilities cannot be completed by the effective date, (ii) he is taking all available steps to safeguard his employees against the hazards covered by the standard, and (iii) he has an effective program for coming into compliance with the standard as quickly as practicable. Any temporary order issued under this paragraph shall prescribe the practices, means, methods, operations, and processes which the employer must adopt and use while the order is in effect and state in detail his program for coming into compliance with the standard. Such a temporary order may be granted only after notice to employees and an opportunity for a hearing: *Provided,* That the Secretary may issue one interim order to be effective until a decision is made on the basis of the hearing. No temporary order may be in effect for longer than the period needed by the employer to achieve compliance with the standard or one year, whichever is shorter, except that such an order may be renewed not more than twice (I) so long as the requirements of this paragraph are met and (II) if an application for renewal is filed at least 90 days prior to the expiration date of the order. No interim renewal of an order may remain in effect for longer than 180 days.

(B) An application for a temporary order under this paragraph (6) shall contain:

(i) a specification of the standard or portion thereof from which the employer seeks a variance,

(ii) a representation by the employer, supported by representations from qualified persons having firsthand knowledge of the facts represented, that he is unable to comply with the standard or portion thereof and a detailed statement of the reasons therefor,

(iii) a statement of the steps he has taken and will take (with specific dates) to protect employees against the hazard covered by the standard,

(iv) a statement of when he expects to be able to comply with the standard and what steps he has taken and what steps he will take (with dates specified) to come into compliance with the standard, and

(v) a certification that he has informed his employees of the application by giving a copy thereof to their authorized representative, posting a statement giving a summary of the application and specifying where a copy may be examined at the place or places where notices to employees are normally posted, and by other appropriate means.

A description of how employees have been informed shall be contained in the certification. The information to employees shall also inform them of their right to petition the Secretary for a hearing.

(C) The Secretary is authorized to grant a variance from any standard or portion thereof whenever he determines, or the Secretary of Health, Education, and Welfare certifies, that such variance is necessary to permit an employer to participate in an experiment approved by him or the Secretary of Health, Education, and Welfare designed to demonstrate or validate new and improved techniques to safeguard the health or safety of workers.

(7) Any standard promulgated under this subsection shall prescribe the use of labels or other appropriate forms of warning as are necessary to insure that employees are apprised of all hazards to which they are exposed, relevant symptoms and appropriate emergency treatment, and proper con-

ditions and precautions of safe use or exposure. Where appropriate, such standard shall also prescribe suitable protective equipment and control or technological procedures to be used in connection with such hazards and shall provide for monitoring or measuring employee exposure at such locations and intervals, and in such manner as may be necessary for the protection of employees. In addition, where appropriate, any such standard shall prescribe the type and frequency of medical examinations or other tests which shall be made available, by the employer or at his cost, to employees exposed to such hazards in order to most effectively determine whether the health of such employees is adversely affected by such exposure. In the event such medical examinations are in the nature of research, as determined by the Secretary of Health, Education, and Welfare, such examinations may be furnished at the expense of the Secretary of Health, Education, and Welfare. The results of such examinations or tests shall be furnished only to the Secretary or the Secretary of Health, Education, and Welfare, and, at the request of the employee, to his physician. The Secretary, in consultation with the Secretary of Health, Education, and Welfare, may by rule promulgated pursuant to section 553 of title 5, United States Code, make appropriate modifications in the foregoing requirements relating to the use of labels or other forms of warning, monitoring or measuring, and medical examinations, as may be warranted by experience, information, or medical or technological developments acquired subsequent to the promulgation of the relevant standard.

(8) Whenever a rule promulgated by the Secretary differs substantially from an existing national consensus standard, the Secretary shall, at the same time, publish in the Federal Register a statement of the reasons why the rule as adopted will better effectuate the purposes of this Act than the national consensus standard.

(c) (1) The Secretary shall provide, without regard to the requirements of chapter 5, title 5, United States Code, for an emergency temporary standard to take immediate effect upon publication in the Federal Register if he determines (A) that employees are exposed to grave danger from exposure to substances or agents determined to be toxic or physically harmful or from new hazards, and (B) that such emergency standard is necessary to protect employees from such danger.

(2) Such standard shall be effective until superseded by a standard promulgated in accordance with the procedures prescribed in paragraph (3) of this subsection.

(3) Upon publication of such standard in the Federal Register the Secretary shall commence a proceeding in accordance with section 6(b) of this Act, and the standard as published shall also serve as a proposed rule for the proceeding. The Secretary shall promulgate a standard under this paragraph no later than six months after publication of the emergency standard as provided in paragraph (2) of this subsection.

(d) Any affected employer may apply to the Secretary for a rule or order for a variance from a standard promulgated under this section. Affected employees shall be given notice of each such application and an opportunity to participate in a hearing. The Secretary shall issue such rule or order if he determines on the record, after opportunity for an inspection where appropriate and a hearing, that the proponent of the variance has demonstrated by a preponderance of the evidence that the conditions, practices, means, methods, operations, or processes used or proposed to be used by an employer will provide employment and places of employment to his employees which are as safe and healthful as those which would prevail if he complied with the standard. The rule or order so issued shall prescribe the conditions the employer must maintain, and the practices, means, methods, operations, and processes which he must adopt and utilize to the extent they differ from the standard in question. Such a rule or order may be modified or revoked upon application by an employer, employees, or by the Secretary on his own motion, in the manner prescribed for its issuance under this subsection at any time after six months from its issuance.

(e) Whenever the Secretary promulgates any standard, makes any rule, order, or decision, grants any exemption or extension of time, or compromises, mitigates, or settles any penalty assessed under this Act, he shall include a statement of the reasons for such action, which shall be published in the Federal Register.

(f) Any person who may be adversely affected by a standard issued under this section may at any time prior to the sixtieth day after such standard is promulgated file a petition challenging the validity of such standard with the United States court of appeals for the circuit wherein such person resides or has his principal place of business, for a judicial review of such standard. A copy of the petition shall be forthwith transmitted by the clerk of the court to the Secretary. The filing of such petition shall not, unless otherwise ordered by the court, operate as a stay of the standard. The determinations of the Secretary shall be conclusive if supported by substantial evidence in the record considered as a whole.

(g) In determining the priority for establishing standards under this section, the Secretary shall give due regard to the urgency of the need for mandatory safety and health standards for particular industries, trades, crafts, occupations, businesses, workplaces or work environments. The Secretary shall also give due regard to the recommendations of the Secretary of Health, Education, and Welfare regarding the need for mandatory standards in determining the priority for establishing such standards.

ADVISORY COMMITTEES; ADMINISTRATION

Sec. 7. (a) (1) There is hereby established a National Advisory Committee on Occupational Safety and Health consisting of twelve members appointed by the Secretary, four of whom are to be designated by the Secretary of Health, Education, and Welfare, without regard to the provisions of title 5, United States Code, governing appointments in the competitive service, and composed of representatives of manangement, labor, occupational safety and occupational health professions, and of the public. The Secretary shall designate one of the public members as Chairman. The members shall be selected upon the basis of their experience and competence in the field of occupational safety and health.

(2) The Committee shall advise, consult with, and make recommendations to the Secretary and the Secretary of Health, Education, and Welfare on matters relating to the administration of the Act. The Committee shall hold no fewer than two meetings during each calendar year. All meetings of the Committee shall be open to the public and a transcript shall be kept and made available for public inspection.

(3) The members of the Committee shall be compensated in accordance with the provisions of section 3109 of title 5, United States Code.

(4) The Secretary shall furnish to the Committee an executive secretary

and such secretarial, clerical, and other services as are deemed necessary to the conduct of its business.

(b) An advisory committee may be appointed by the Secretary to assist him in his standard-setting functions under section 6 of this Act. Each such committee shall consist of not more than fifteen members and shall include as a member one or more designees of the Secretary of Health, Education, and Welfare, and shall include among its members an equal number of persons qualified by experience and affiliation to present the viewpoint of the employers involved, and of persons similarly qualified to present the viewpoint of the workers involved, as well as one or more representatives of health and safety agencies of the States. An advisory committee may also include such other persons as the Secretary may appoint who are qualified by knowledge and experience to make a useful contribution to the work of such committee, including one or more representatives of professional organizations of technicians or professionals specializing in occupational safety or health, and one or more representatives of nationally recognized standards-producing organizations, but the number of persons so appointed to any such advisory committee shall not exceed the number appointed to such committee as representatives of Federal and State agencies. Persons appointed to advisory committees from private life shall be compensated in the same manner as consultants or experts under section 3109 of title 5, United States Code. The Secretary shall pay to any State which is the employer of a member of such a committee who is a representative of the health or safety agency of that State, reimbursement sufficient to cover the actual cost to the State resulting from such representative's membership on such committee. Any meeting of such committee shall be open to the public and an accurate record shall be kept and made available to the public. No member of such committee (other than representatives of employers and employees) shall have an economic interest in any proposed rule.

(c) In carrying out his responsibilities under this Act, the Secretary is authorized to—

(1) use, with the consent of any Federal agency, the services, facilities, and personnel of such agency, with or without reimbursement, and with the consent of any State or political subdivision thereof, accept and use the services, facilities, and personnel of any agency of such State or subdivision with reimbursement; and

(2) employ experts and consultants or organizations thereof as authorized by section 3109 of title 5, United States Code, except that contracts for such employment may be renewed annually; compensate individuals so employed at rates not in excess of the rate specified at the time of service for grade GS–18 under section 5332 of title 5, United States Code, including traveltime, and allow them while away from their homes or regular places of business, travel expenses (including per diem in lieu of subsistence) as authorized by section 5703 of title 5, United States Code, for persons in the Government service employed intermittently, while so employed.

INSPECTIONS, INVESTIGATIONS, AND RECORDKEEPING

SEC. 8. (a) In order to carry out the purposes of this Act, the Secretary, upon presenting appropriate credentials to the owner, operator, or agent in charge, is authorized—

(1) to enter without delay and at reasonable times any factory, plant,

establishment, construction site, or other area, workplace or environment where work is performed by an employee of an employer; and

(2) to inspect and investigate during regular working hours and at other reasonable times, and within reasonable limits and in a reasonable manner, any such place of employment and all pertinent conditions, structures, machines, apparatus, devices, equipment, and materials therein, and to question privately any such employer, owner, operator, agent or employee.

(b) In making his inspections and investigations under this Act the Secretary may require the attendance and testimony of witnesses and the production of evidence under oath. Witnesses shall be paid the same fees and mileage that are paid witnesses in the courts of the United States. In case of a contumacy, failure, or refusal of any person to obey such an order, any district court of the United States or the United States courts of any territory or possession, within the jurisdiction of which such person is found, or resides or transacts business, upon the application by the Secretary, shall have jurisdiction to issue to such person an order requiring such person to appear to produce evidence if, as, and when so ordered, and to give testimony relating to the matter under investigation or in question, and any failure to obey such order of the court may be punished by said court as a contempt thereof.

(c) (1) Each employer shall make, keep and preserve, and make available to the Secretary or the Secretary of Health, Education, and Welfare, such records regarding his activities relating to this Act as the Secretary, in cooperation with the Secretary of Health, Education, and Welfare, may prescribe by regulation as necessary or appropriate for the enforcement of this Act or for developing information regarding the causes and prevention of occupational accidents and illnesses. In order to carry out the provisions of this paragraph such regulations may include provisions requiring employers to conduct periodic inspections. The Secretary shall also issue regulations requiring that employers, through posting of notices or other appropriate means, keep their employees informed of their protections and obligations under this Act, including the provisions of applicable standards.

(2) The Secretary, in cooperation with the Secretary of Health, Education, and Welfare, shall prescribe regulations requiring employers to maintain accurate records of, and to make periodic reports on, work-related deaths, injuries and illnesses other than minor injuries requiring only first aid treatment and which do not involve medical treatment, loss of consciousness, restriction of work or motion, or transfer to another job.

(3) The Secretary, in cooperation with the Secretary of Health, Education, and Welfare, shall issue regulations requiring employers to maintain accurate records of employee exposures to potentially toxic materials or harmful physical agents which are required to be monitored or measured under section 6. Such regulations shall provide employees or their representatives with an opportunity to observe such monitoring or measuring, and to have access to the records thereof. Such regulations shall also make appropriate provision for each employee or former employee to have access to such records as will indicate his own exposure to toxic materials or harmful physical agents. Each employer shall promptly notify any employee who has been or is being exposed to toxic materials or harmful physical agents in concentrations or at levels which exceed those prescribed by an applicable occupa-

tional safety and health standard promulgated under section 6, and shall inform any employee who is being thus exposed of the corrective action being taken.

(d) Any information obtained by the Secretary, the Secretary of Health, Education, and Welfare, or a State agency under this Act shall be obtained with a minimum burden upon employers, especially those operating small businesses. Unnecessary duplication of efforts in obtaining information shall be reduced to the maximum extent feasible.

(e) Subject to regulations issued by the Secretary, a representative of the employer and a representative authorized by his employees shall be given an opportunity to accompany the Secretary or his authorized representative during the physical inspection of any workplace under subsection (a) for the purpose of aiding such inspection. Where there is no authorized employee representative, the Secretary or his authorized representative shall consult with a reasonable number of employees concerning matters of health and safety in the workplace.

(f) (1) Any employees or representative of employees who believe that a violation of a safety or health standard exists that threatens physical harm, or that an imminent danger exists, may request an inspection by giving notice to the Secretary or his authorized representative of such violation or danger. Any such notice shall be reduced to writing, shall set forth with reasonable particularity the grounds for the notice, and shall be signed by the employees or representative of employees, and a copy shall be provided the employer or his agent no later than at the time of inspection, except that, upon the request of the person giving such notice, his name and the names of individual employees referred to therein shall not appear in such copy or on any record published, released, or made available pursuant to subsection (g·) of this section. If upon receipt of such notification the Secretary determines there are reasonable grounds to believe that such violation or danger exists, he shall make a special inspection in accordance with the provisions of this section as soon as practicable, to determine if such violation or danger exists. If the Secretary determines there are no reasonable grounds to believe that a violation or danger exists he shall notify the employees or representative of the employees in writing of such determination.

(2) Prior to or during any inspection of a workplace, any employees or representative of employees employed in such workplace may notify the Secretary or any representative of the Secretary responsible for conducting the inspection, in writing, of any violation of this Act which they have reason to believe exists in such workplace. The Secretary shall, by regulation, establish procedures for informal review of any refusal by a representative of the Secretary to issue a citation with respect to any such alleged violation and shall furnish the employees or representative of employees requesting such review a written statement of the reasons for the Secretary's final disposition of the case.

(g) (1) The Secretary and Secretary of Health, Education, and Welfare are authorized to compile, analyze, and publish, either in summary or detailed form, all reports or information obtained under this section.

(2) The Secretary and the Secretary of Health, Education, and Welfare shall each prescribe such rules and regulations as he may deem necessary to carry out their responsibilities under this Act, including rules and regulations dealing with the inspection of an employer's establishment.

CITATIONS

SEC. 9. (a) If, upon inspection or investigation, the Secretary or his authorized representative believes that an employer has violated a requirement of section 5 of this Act, of any standard, rule or order promulgated pursuant to section 6 of this Act, or of any regulations prescribed pursuant to this Act, he shall with reasonable promptness issue a citation to the employer. Each citation shall be in writing and shall describe with particularity the nature of the violation, including a reference to the provision of the Act, standard, rule, regulation, or order alleged to have been violated. In addition, the citation shall fix a reasonable time for the abatement of the violation. The Secretary may prescribe procedures for the issuance of a notice in lieu of a citation with respect to de minimis violations which have no direct or immediate relationship to safety or health.

(b) Each citation issued under this section, or a copy or copies thereof, shall be prominently posted, as prescribed in regulations issued by the Secretary, at or near each place a violation referred to in the citation occurred.

(c) No citation may be issued under this section after the expiration of six months following the occurrence of any violation.

PROCEDURE FOR ENFORCEMENT

SEC. 10. (a) If, after an inspection or investigation, the Secretary issues a citation under section 9(a), he shall, within a reasonable time after the termination of such inspection or investigation, notify the employer by certified mail of the penalty, if any, proposed to be assessed under section 17 and that the employer has fifteen working days within which to notify the Secretary that he wishes to contest the citation or proposed assessment of penalty. If, within fifteen working days from the receipt of the notice issued by the Secretary the employer fails to notify the Secretary that he intends to contest the citation or proposed assessment of penalty, and no notice is filed by any employee or representative of employees under subsection (c) within such time, the citation and the assessment, as proposed, shall be deemed a final order of the Commission and not subject to review by any court or agency.

(b) If the Secretary has reason to believe that an employer has failed to correct a violation for which a citation has been issued within the period permitted for its correction (which period shall not begin to run until the entry of a final order by the Commission in the case of any review proceedings under this section initiated by the employer in good faith and not solely for delay or avoidance of penalties), the Secretary shall notify the employer by certified mail of such failure and of the penalty proposed to be assessed under section 17 by reason of such failure, and that the employer has fifteen working days within which to notify the Secretary that he wishes to contest the Secretary's notification or the proposed assessment of penalty. If, within fifteen working days from the receipt of notification issued by the Secretary, the employer fails to notify the Secretary that he intends to contest the notification or proposed assessment of penalty, the notification and assessment, as proposed, shall be deemed a final order of the Commission and not subject to review by any court or agency.

(c) If an employer notifies the Secretary that he intends to contest a citation issued under section 9(a) or notification issued under subsection (a) or (b) of this section, or if, within fifteen working days of the issuance of a citation under section 9(a), any employee or representative of employees files a notice with the Secretary alleging that the period of time fixed in the citation for the abatement of the violation is unreasonable, the Secretary shall immediately advise the Commission of such notification, and the Commission shall afford an opportunity for a hearing (in accordance with section 554 of title 5, United States Code, but without regard to subsection (a) (3) of such section). The Commission shall thereafter issue an order, based on findings of fact, affirming, modifying, or vacating the Secretary's citation or proposed penalty, or directing other appropriate relief, and such order shall become final thirty days after its issuance. Upon a showing by an employer of a good faith effort to comply with the abatement requirements of a citation, and that abatement has not been completed because of factors beyond his reasonable control, the Secretary, after an opportunity for a hearing as provided in this subsection, shall issue an order affirming or modifying the abatement requirements in such citation. The rules of procedure prescribed by the Commission shall provide affected employees or representatives of affected employees an opportunity to participate as parties to hearings under this subsection.

JUDICIAL REVIEW

SEC. 11. (a) Any person adversely affected or aggrieved by an order of the Commission issued under subsection (c) of section 10 may obtain a review of such order in any United States court of appeals for the circuit in which the violation is alleged to have occurred or where the employer has its principal office, or in the Court of Appeals for the District of Columbia Circuit, by filing in such court within sixty days following the issuance of such order a written petition praying that the order be modified or set aside. A copy of such petition shall be forthwith transmitted by the clerk of the court to the Commission and to the other parties, and thereupon the Commission shall file in the court the record in the proceeding as provided in section 2112 of title 28, United States Code. Upon such filing, the court shall have jurisdiction of the proceeding and of the question determined therein, and shall have power to grant such temporary relief or restraining order as it deems just and proper, and to make and enter upon the pleadings, testimony, and proceedings set forth in such record a decree affirming, modifying, or setting aside in whole or in part, the order of the Commission and enforcing the same to the extent that such order is affirmed or modified. The commencement of proceedings under this subsection shall not, unless ordered by the court, operate as a stay of the order of the Commission. No objection that has not been urged before the Commission shall be considered by the court, unless the failure or neglect to urge such objection shall be excused because of extraordinary circumstances. The findings of the Commission with respect to questions of fact, if supported by substantial evidence on the record considered as a whole, shall be conclusive. If any party shall apply to the court for leave to adduce additional evidence and shall show to the satisfaction of the court that such additional evidence is material and that there were reasonable grounds for the failure to adduce such evidence in the hearing before the Commission, the court may order such additional evidence to be taken before the Commission and to be made a part of the record. The Commission may modify its findings as to the facts, or make new findings, by reason of additional evidence so taken and filed, and it shall file such modified or new findings, which findings with respect to questions of fact, if supported by substantial evidence on the record considered as a whole, shall be conclusive, and its recommendations, if any, for the modification or setting aside of its original order. Upon the filing of the record with it, the jurisdiction of the court shall be exclusive and its judgment and decree shall be final, except that the same shall be subject to review by the Supreme Court of the United States, as provided in section 1254 of title 28, United States Code. Petitions filed under this subsection shall be heard expeditiously.

(b) The Secretary may also obtain review or enforcement of any final order of the Commission by filing a petition for such relief in the United States court of appeals for the circuit in which the alleged violation occurred or in which the employer has its principal office, and the provisions of subsection (a) shall govern such proceedings to the extent applicable. If no petition for review, as provided in subsection (a), is filed within sixty days after service of the Commission's order, the Commission's findings of fact and order shall be conclusive in connection with any petition for enforcement which is filed by the Secretary after the expiration of such sixty-day period. In any such case, as well as in the case of a noncontested citation or notification by the Secretary which has become a final order of the Commission under subsection (a) or (b) of section 10, the clerk of the court, unless otherwise ordered by the court, shall forthwith enter a decree enforcing the order and shall transmit a copy of such decree to the Secretary and the employer named in the petition. In any contempt proceeding brought to enforce a decree of a court of appeals entered pursuant to this subsection or subsection (a), the court of appeals may assess the penalties provided in section 17, in addition to invoking any other available remedies.

(c) (1) No person shall discharge or in any manner discriminate against any employee because such employee has filed any complaint or instituted or caused to be instituted any proceeding under or related to this Act or has testified or is about to testify in any such proceeding or because of the exercise by such employee on behalf of himself or others of any right afforded by this Act.

(2) Any employee who believes that he has been discharged or otherwise discriminated against by any person in violation of this subsection may, within thirty days after such violation occurs, file a complaint with the Secretary alleging such discrimination. Upon receipt of such complaint, the Secretary shall cause such investigation to be made as he deems appropriate. If upon such investigation, the Secretary determines that the provisions of this subsection have been violated, he shall bring an action in any appropriate United States district court against such person. In any such action the United States district courts shall have jurisdiction, for cause shown to restrain violations of paragraph (1) of this subsection and order all appropriate relief including rehiring or reinstatement of the employee to his former position with back pay.

(3) Within 90 days of the receipt of a complaint filed under this subsection the Secretary shall notify the complainant of his determination under paragraph 2 of this subsection.

THE OCCUPATIONAL SAFETY AND HEALTH REVIEW COMMISSION

SEC. 12. (a) The Occupational Safety and Health Review Commission is hereby established. The Commission shall be composed of three members who shall be appointed by the President, by and with the advice and consent of the Senate, from among persons who by reason of training, education, or experience are qualified to carry out the functions of the Commission under this Act. The President shall designate one of the members of the Commission to serve as Chairman.

(b) The terms of members of the Commission shall be six years except that (1) the members of the Commission first taking office shall serve, as designated by the President at the time of appointment, one for a term of two years, one for a term of four years, and one for a term of six years, and (2) a vacancy caused by the death, resignation, or removal of a member prior to the expiration of the term for which he was appointed shall be filled only for the remainder of such unexpired term. A member of the Commission may be removed by the President for inefficiency, neglect of duty, or malfeasance in office.

(c) (1) Section 5314 of title 5, United States Code, is amended by adding at the end thereof the following new paragraph:

"(57) Chairman, Occupational Safety and Health Review Commission."

(2) Section 5315 of title 5, United States Code, is amended by adding at the end thereof the following new paragraph:

"(94) Members, Occupational Safety and Health Review Commission."

(d) The principal office of the Commission shall be in the District of Columbia. Whenever the Commission deems that the convenience of the public or of the parties may be promoted, or delay or expense may be minimized, it may hold hearings or conduct other proceedings at any other place.

(e) The Chairman shall be responsible on behalf of the Commission for the administrative operations of the Commission and shall appoint such hearing examiners and other employees as he deems necessary to assist in the performance of the Commission's functions and to fix their compensation in accordance with the provisions of chapter 51 and subchapter III of chapter 53 of title 5, United States Code, relating to classification and General Schedule pay rates: *Provided*, That assignment, removal and compensation of hearing examiners shall be in accordance with sections 3105, 3344, 5362, and 7521 of title 5, United States Code.

(f) For the purpose of carrying out its functions under this Act, two members of the Commission shall constitute a quorum and official action can be taken only on the affirmative vote of at least two members.

(g) Every official act of the Commission shall be entered of record, and its hearings and records shall be open to the public. The Commission is authorized to make such rules as are necessary for the orderly transaction of its proceedings. Unless the Commission has adopted a different rule, its proceedings shall be in accordance with the Federal Rules of Civil Procedure.

(h) The Commission may order testimony to be taken by deposition in any proceedings pending before it at any state of such proceeding. Any person may be compelled to appear and depose, and to produce books, papers, or documents, in the same manner as witnesses may be compelled to appear and testify and produce like documentary evidence before the Commission. Witnesses whose depositions are taken under this subsection, and the persons taking such depositions, shall be entitled to the same fees as are paid for like services in the courts of the United States.

(i) For the purpose of any proceeding before the Commission, the provisions of section 11 of the National Labor Relations Act (29 U.S.C. 161) are hereby made applicable to the jurisdiction and powers of the Commission.

(j) A hearing examiner appointed by the Commission shall hear, and make a determination upon, any proceeding instituted before the Commission and any motion in connection therewith, assigned to such hearing examiner by the Chairman of the Commission, and shall make a report of any such determination which constitutes his final disposition of the proceedings. The report of the hearing examiner shall become the final order of the Commission within thirty days after such report by the hearing examiner, unless within such period any Commission member has directed that such report shall be reviewed by the Commission.

(k) Except as otherwise provided in this Act, the hearing examiners shall be subject to the laws governing employees in the classified civil service, except that appointments shall be made without regard to section 5108 of title 5, United States Code. Each hearing examiner shall receive compensation at a rate not less than that prescribed for GS–16 under section 5332 of title 5, United States Code.

PROCEDURES TO COUNTERACT IMMINENT DANGERS

Sec. 13. (a) The United States district courts shall have jurisdiction, upon petition of the Secretary, to restrain any conditions or practices in any place of employment which are such that a danger exists which could reasonably be expected to cause death or serious physical harm immediately or before the imminence of such danger can be eliminated through the enforcement procedures otherwise provided by this Act. Any order issued under this section may require such steps to be taken as may be necessary to avoid, correct, or remove such imminent danger and prohibit the employment or presence of any individual in locations or under conditions where such imminent danger exists, except individuals whose presence is necessary to avoid, correct, or remove such imminent danger or to maintain the capacity of a continuous process operation to resume normal operations without a complete cessation of operations, or where a cessation of operations is necessary, to permit such to be accomplished in a safe and orderly manner.

(b) Upon the filing of any such petition the district court shall have jurisdiction to grant such injunctive relief or temporary restraining order pending the outcome of an enforcement proceeding pursuant to this Act. The proceeding shall be as provided by Rule 65 of the Federal Rules, Civil Procedure, except that no temporary restraining order issued without notice shall be effective for a period longer than five days.

(c) Whenever and as soon as an inspector concludes that conditions or practices described in subsection (a) exist in any place of employment, he shall inform the affected employees and employers of the danger and that he is recommending to the Secretary that relief be sought.

(d) If the Secretary arbitrarily or capriciously fails to seek relief under this section, any employee who may be injured by reason of such failure, or the representative of such employees, might bring an action against the Secretary in the United States district court for the district in which the imminent danger is alleged to exist or the employer has its principal office, or for the District of Columbia, for a writ of mandamus to compel the Secretary to seek such an order and for such further relief as may be appropriate.

REPRESENTATION IN CIVIL LITIGATION

SEC. 14. Except as provided in section 518(a) of title 28, United States Code, relating to litigation before the Supreme Court, the Solicitor of Labor may appear for and represent the Secretary in any civil litigation brought under this Act but all such litigation shall be subject to the direction and control of the Attorney General.

CONFIDENTIALITY OF TRADE SECRETS

SEC. 15. All information reported to or otherwise obtained by the Secretary or his representative in connection with any inspection or proceeding under this Act which contains or which might reveal a trade secret referred to in section 1905 of title 18 of the United States Code shall be considered confidential for the purpose of that section, except that such information may be disclosed to other officers or employees concerned with carrying out this Act or when relevant in any proceeding under this Act. In any such proceeding the Secretary, the Commission, or the court shall issue such orders as may be appropriate to protect the confidentiality of trade secrets.

VARIATIONS, TOLERANCES, AND EXEMPTIONS

SEC. 16. The Secretary, on the record, after notice and opportunity for a hearing may provide such reasonable limitations and may make such rules and regulations allowing reasonable variations, tolerances, and exemptions to and from any or all provisions of this Act as he may find necessary and proper to avoid serious impairment of the national defense. Such action shall not be in effect for more than six months without notification to affected employees and an opportunity being afforded for a hearing.

PENALTIES

SEC. 17. (a) Any employer who willfully or repeatedly violates the requirements of section 5 of this Act, any standard, rule, or order promulgated pursuant to section 6 of this Act, or regulations prescribed pursuant to this Act, may be assessed a civil penalty of not more than $10,000 for each violation.

(b) Any employer who has received a citation for a serious violation of the requirements of section 5 of this Act, of any standard, rule, or order promulgated pursuant to section 6 of this Act, or of any regulations prescribed pursuant to this Act, shall be assessed a civil penalty of up to $1,000 for each such violation.

(c) Any employer who has received a citation for a violation of the requirements of section 5 of this Act, of any standard, rule, or order promulgated pursuant to section 6 of this Act, or of regulations prescribed pursuant to this Act, and such violation is specifically determined not to be of a serious nature, may be assessed a civil penalty of up to $1,000 for each such violation.

(d) Any employer who fails to correct a violation for which a citation has been issued under section 9(a) within the period permitted for its correction (which period shall not begin to run until the date of the final order of the Commission in the case of any review proceeding under section 10 initiated by the employer in good faith and not solely for delay or avoidance of penalties), may be assessed a civil penalty of not more than $1,000 for each day during which such failure or violation continues.

(e) Any employer who willfully violates any standard, rule, or order promulgated pursuant to section 6 of this Act, or of any regulations prescribed pursuant to this Act, and that violation caused death to any employee, shall, upon conviction, be punished by a fine of not more than $10,000 or by imprisonment for not more than six months, or by both; except that if the conviction is for a violation committed after a first conviction of such person, punishment shall be by a fine of not more than $20,000 or by imprisonment for not more than one year, or by both.

(f) Any person who gives advance notice of any inspection to be conducted under this Act, without authority from the Secretary or his designees, shall, upon conviction, be punished by a fine of not more than $1,000 or by imprisonment for not more than six months, or by both.

(g) Whoever knowingly makes any false statement, representation, or certification in any application, record, report, plan, or other document filed or required to be maintained pursuant to this Act shall, upon conviction, be punished by a fine of not more than $10,000, or by imprisonment for not more than six months, or by both.

(h) (1) Section 1114 of title 18, United States Code, is hereby amended by striking out "designated by the Secretary of Health, Education, and Welfare to conduct investigations, or inspections under the Federal Food, Drug, and Cosmetic Act" and inserting in lieu thereof "or of the Department of Labor assigned to perform investigative, inspection, or law enforcement functions".

(2) Notwithstanding the provisions of sections 1111 and 1114 of title 18, United States Code, whoever, in violation of the provisions of section 1114 of such title, kills a person while engaged in or on account of the performance of investigative, inspection, or law enforcement functions added to such section 1114 by paragraph (1) of this subsection, and who would otherwise be subject to the penalty provisions of such section 1111, shall be punished by imprisonment for any term of years or for life.

(i) Any employer who violates any of the posting requirements, as prescribed under the provisions of this Act, shall be assessed a civil penalty of up to $1,000 for each violation.

(j) The Commission shall have authority to assess all civil penalties provided in this section, giving due consideration to the appropriateness of the penalty with respect to the size of the business of employer being charged, the gravity of the violation, the good faith of the employer, and the history of previous violations.

(k) For purposes of this section, a serious violation shall be deemed to exist in a place of employment if there is a substantial probability that death or serious physical harm could result from a condition which exists, or from one or more practices, means, methods, operations, or processes which have been adopted or are in use, in such place of employment unless the employer did not, and could not with the exercise of reasonable diligence, know of the presence of the violation.

(1) Civil penalties owed under this Act shall be paid to the Secretary for deposit into the Treasury of the United States and shall accrue to the United States and may be recovered in a civil action in the name of the United States brought in the United States district court for the district where the violation is alleged to have occurred or where the employer has its principal office.

STATE JURISDICTION AND STATE PLANS

Sec. 18. (a) Nothing in this Act shall prevent any State agency or court from asserting jurisdiction under State law over any occupational safety or health issue with respect to which no standard is in effect under section 6.

(b) Any State which, at any time, desires to assume responsibility for development and enforcement therein of occupational safety and health standards relating to any occupational safety or health issue with respect to which a Federal standard has been promulgated under section 6 shall submit a State plan for the development of such standards and their enforcement.

(c) The Secretary shall approve the plan submitted by a State under subsection (b), or any modification thereof, if such plan in his judgment—

(1) designates a State agency or agencies as the agency or agencies responsible for administering the plan throughout the State,

(2) provides for the development and enforcement of safety and health standards relating to one or more safety or health issues, which standards (and the enforcement of which standards) are or will be at least as effective in providing safe and healthful employment and places of employment as the standards promulgated under section 6 which relate to the same issues, and which standards, when applicable to products which are distributed or used in interstate commerce, are required by compelling local conditions and do not unduly burden interstate commerce,

(3) provides for a right of entry and inspection of all workplaces subject to the Act which is at least as effective as that provided in section 8, and includes a prohibition on advance notice of inspections,

(4) contains satisfactory assurances that such agency or agencies have or will have the legal authority and qualified personnel necessary for the enforcement of such standards,

(5) gives satisfactory assurances that such State will devote adequate funds to the administration and enforcement of such standards,

(6) contains satisfactory assurances that such State will, to the extent permitted by its law, establish and maintain an effective and comprehensive occupational safety and health program applicable to all employees of public agencies of the State and its political subdivisions, which program is as effective as the standards contained in an approved plan,

(7) requires employers in the State to make reports to the Secretary in the same manner and to the same extent as if the plan were not in effect, and

(8) provides that the State agency will make such reports to the Secretary in such form and containing such information, as the Secretary shall from time to time require.

(d) If the Secretary rejects a plan submitted under subsection (b), he shall afford the State submitting the plan due notice and opportunity for a hearing before so doing.

(e) After the Secretary approves a State plan submitted under subsection (b), he may, but shall not be required to, exercise his authority under sections 8, 9, 10, 13, and 17 with respect to comparable standards promulgated under section 6, for the period specified in the next sentence. The Secretary may exercise the authority referred to above until he determines, on the basis of actual operations under the State plan, that the criteria set forth in subsection (c) are being applied, but he shall not make such determination for at least three years after the plan's approval under subsection (c). Upon making the determination referred to in the preceding sentence, the provisions of sections 5(a) (2), 8 (except for the purpose of carrying out subsection (f) of this section), 9, 10, 13, and 17, and standards promulgated under section 6 of this Act, shall not apply with respect to any occupational safety or health issues covered under the plan, but the Secretary may retain jurisdiction under the above provisions in any proceeding commenced under section 9 or 10 before the date of determination.

(f) The Secretary shall, on the basis of reports submitted by the State agency and his own inspections make a continuing evaluation of the manner in which each State having a plan approved under this section is carrying out such plan. Whenever the Secretary finds, after affording due notice and opportunity for a hearing, that in the administration of the State plan there is a failure to comply substantially with any provision of the State plan (or any assurance contained therein), he shall notify the State agency of his withdrawal of approval of such plan and upon receipt of such notice such plan shall cease to be in effect, but the State may retain jurisdiction in any case commenced before the withdrawal of the plan in order to enforce standards under the plan whenever the issues involved do not relate to the reasons for the withdrawal of the plan.

(g) The State may obtain a review of a decision of the Secretary withdrawing approval of or rejecting its plan by the United States court of appeals for the circuit in which the State is located by filing in such court within thirty days following receipt of notice of such decision a petition to modify or set aside in whole or in part the action of the Secretary. A copy of such petition shall forthwith be served upon the Secretary, and thereupon the Secretary shall certify and file in the court the record upon which the decision complained of was issued as provided in section 2112 of title 28, United States Code. Unless the court finds that the Secretary's decision in rejecting a proposed State plan or withdrawing his approval of such a plan is not supported by substantial evidence the court shall affirm the Secretary's decision. The judgment of the court shall be subject to review by the Supreme Court of the United States upon certiorari or certification as provided in section 1254 of title 28, United States Code.

(h) The Secretary may enter into an agreement with a State under which the State will be permitted to continue to enforce one or more occupational health and safety standards in effect in such State until final action is taken by the Secretary with respect to a plan submitted by a State under subsection (b) of this section, or two years from the date of enactment of this Act, whichever is earlier.

FEDERAL AGENCY SAFETY PROGRAMS AND RESPONSIBILITIES

SEC. 19. (a) It shall be the responsibility of the head of each Federal agency to establish and maintain an effective and comprehensive occupational safety and health program which is consistent with the standards promulgated under section 6. The head of each agency shall (after consultation with representatives of the employees thereof)—

(1) provide safe and healthful places and conditions of employment, consistent with the standards set under section 6;

(2) acquire, maintain, and require the use of safety equipment, per-

sonal protective equipment, and devices reasonably necessary to protect employees;

(3) keep adequate records of all occupational accidents and illnesses for proper evaluation and necessary corrective action;

(4) consult with the Secretary with regard to the adequacy as to form and content of records kept pursuant to subsection (a) (3) of this section; and

(5) make an annual report to the Secretary with respect to occupational accidents and injuries and the agency's program under this section. Such report shall include any report submitted under section 7902(e) (2) of title 5, United States Code.

(b) The Secretary shall report to the President a summary or digest of reports submitted to him under subsection (a) (5) of this section, together with his evaluations of and recommendations derived from such reports. The President shall transmit annually to the Senate and the House of Representatives a report of the activities of Federal agencies under this section.

(c) Section 7902(c) (1) of title 5, United States Code, is amended by inserting after "agencies" the following: "and of labor organizations representing employees".

(d) The Secretary shall have access to records and reports kept and filed by Federal agencies pursuant to subsections (a) (3) and (5) of this section unless those records and reports are specifically required by Executive order to be kept secret in the interest of the national defense or foreign policy, in which case the Secretary shall have access to such information as will not jeopardize national defense or foreign policy.

RESEARCH AND RELATED ACTIVITIES

Sec. 20. (a) (1) The Sercretary of Health, Education, and Welfare, after consultation with the Secretary and with other appropriate Federal departments or agencies, shall conduct (directly or by grants or contracts) research, experiments, and demonstrations relating to occupational safety and health, including studies of psychological factors involved, and relating to innovative methods, techniques, and approaches for dealing with occupational safety and health problems.

(2) The Secretary of Health, Education, and Welfare shall from time to time consult with the Secretary in order to develop specific plans for such research, demonstrations, and experiments as are necessary to produce criteria, including criteria identifying toxic substances, enabling the Secretary to meet his responsibility for the formulation of safety and health standards under this Act; and the Secretary of Health, Education, and Welfare, on the basis of such research, demonstrations, and experiments and any other information available to him, shall develop and publish at least annually such criteria as will effectuate the purposes of this Act.

(3) The Secretary of Health, Education, and Welfare, on the basis of such research, demonstrations, and experiments, and any other information available to him, shall develop criteria dealing with toxic materials and harmful physical agents and substances which will describe exposure levels that are safe for various periods of employment, including but not limited to the exposure levels at which no employee will suffer impaired health or functional capacities or diminished life expectancy as a result of his work experience.

(4) The Secretary of Health, Education, and Welfare shall also conduct

special research, experiments, and demonstrations relating to occupational safety and health as are necessary to explore new problems, including those created by new technology in occupational safety and health, which may require ameliorative action beyond that which is otherwise provided for in the operating provisions of this Act. The Secretary of Health, Education, and Welfare shall also conduct research into the motivational and behavioral factors relating to the field of occupational safety and health.

(5) The Secretary of Health, Education, and Welfare, in order to comply with his responsibilities under paragraph (2), and in order to develop needed information regarding potentially toxic substances or harmful physical agents, may prescribe regulations requiring employers to measure, record, and make reports on the exposure of employees to substances or physical agents which the Secretary of Health, Education, and Welfare reasonably believes may endanger the health or safety of employees. The Secretary of Health, Education, and Welfare also is authorized to establish such programs of medical examinations and tests as may be necessary for determining the incidence of occupational illnesses and the susceptibility of employees to such illnesses. Nothing in this or any other provision of this Act shall be deemed to authorize or require medical examination, immunization, or treatment for those who object thereto on religious grounds, except where such is necessary for the protection of the health or safety of others. Upon the request of any employer who is required to measure and record exposure of employees to substances or physical agents as provided under this subsection, the Secretary of Health, Education, and Welfare shall furnish full financial or other assistance to such employer for the purpose of defraying any additional expense incurred by him in carrying out the measuring and recording as provided in this subsection.

(6) The Secretary of Health, Education, and Welfare shall publish within six months of enactment of this Act and thereafter as needed but at least annually a list of all known toxic substances by generic family or other useful grouping, and the concentrations at which such toxicity is known to occur. He shall determine following a written request by any employer or authorized representative of employees, specifying with reasonable particularity the grounds on which the request is made, whether any substance normally found in the place of employment has potentially toxic effects in such concentrations as used or found; and shall submit such determination both to employers and affected employees as soon as possible. If the Secretary of Health, Education, and Welfare determines that any substance is potentially toxic at the concentrations in which it is used or found in a place of employment, and such substance is not covered by an occupational safety or health standard promulgated under section 6, the Secretary of Health, Education, and Welfare shall immediately submit such determination to the Secretary, together with all pertinent criteria.

(7) Within two years of enactment of this Act, and annually thereafter the Secretary of Health, Education, and Welfare shall conduct and publish industrywide studies of the effect of chronic or low-level exposure to industrial materials, processes, and stresses on the potential for illness, disease, or loss of functional capacity in aging adults.

(b) The Secretary of Health, Education, and Welfare is authorized to make inspections and question employers and employees as provided in section 8 of this Act in order to carry out his functions and responsibilities under this section.

(c) The Secretary is authorized to enter into contracts, agreements, or other arrangements with appropriate public agencies or private organizations for the purpose of conducting studies relating to his responsibilities under this Act. In carrying out his responsibilities under this subsection, the Secretary shall cooperate with the Secretary of Health, Education, and Welfare in order to avoid any duplication of efforts under this section.

(d) Information obtained by the Secretary and the Secretary of Health, Education, and Welfare under this section shall be disseminated by the Secretary to employers and employees and organizations thereof.

(e) The functions of the Secretary of Health, Education, and Welfare under this Act shall, to the extent feasible, be delegated to the Director of the National Institute for Occupational Safety and Health established by section 22 of this Act.

TRAINING AND EMPLOYEE EDUCATION

SEC. 21. (a) The Secretary of Health, Education, and Welfare, after consultation with the Secretary and with other appropriate Federal departments and agencies, shall conduct, directly or by grants or contracts (1) education programs to provide an adequate supply of qualified personnel to carry out the purposes of this Act, and (2) informational programs on the importance of and proper use of adequate safety and health equipment.

(b) The Secretary is also authorized to conduct, directly or by grants or contracts, short-term training of personnel engaged in work related to his responsibilities under this Act.

(c) The Secretary, in consultation with the Secretary of Health, Education, and Welfare, shall (1) provide for the establishment and supervision of programs for the education and training of employers and employees in the recognition, avoidance, and prevention of unsafe or unhealthful working conditions in employments covered by this Act, and (2) consult with and advise employers and employees, and organizations representing employers and employees as to effective means of preventing occupational injuries and illnesses.

NATIONAL INSTITUTE FOR OCCUPATIONAL SAFETY AND HEALTH

SEC. 22. (a) It is the purpose of this section to establish a National Institute for Occupational Safety and Health in the Department of Health, Education, and Welfare in order to carry out the policy set forth in section 2 of this Act and to perform the functions of the Secretary of Health, Education, and Welfare under sections 20 and 21 of this Act.

(b) There is hereby established in the Department of Health, Education, and Welfare a National Institute for Occupational Safety and Health. The Institute shall be headed by a Director who shall be appointed by the Secretary of Health, Education, and Welfare, and who shall serve for a term of six years unless previously removed by the Secretary of Health, Education, and Welfare.

(c) The Institute is authorized to—

(1) develop and establish recommended occupational safety and health standards; and

(2) perform all functions of the Secretary of Health, Education, and Welfare under sections 20 and 21 of this Act.

(d) Upon his own initiative, or upon the request of the Secretary or the

Secretary of Health, Education, and Welfare, the Director is authorized (1) to conduct such research and experimental programs as he determines are necessary for the development of criteria for new and improved occupational safety and health standards, and (2) after consideration of the results of such research and experimental programs make recommendations concerning new or improved occupational safety and health standards. Any occupational safety and health standard recommended pursuant to this section shall immediately be forwarded to the Secretary of Labor, and to the Secretary of Health, Education, and Welfare.

(e) In addition to any authority vested in the Institute by other provisions of this section, the Director, in carrying out the functions of the Institute, is authorized to—

(1) prescribe such regulations as he deems necessary governing the manner in which its functions shall be carried out;

(2) receive money and other property donated, bequeathed, or devised, without condition or restriction other than that it be used for the purposes of the Institute and to use, sell, or otherwise dispose of such property for the purpose of carrying out its functions;

(3) receive (and use, sell, or otherwise dispose of, in accordance with paragraph (2)), money and other property donated, bequeathed, or devised to the Institute with a condition or restriction, including a condition that the Institute use other funds of the Institute for the purposes of the gift;

(4) in accordance with the civil service laws, appoint and fix the compensation of such personnel as may be necessary to carry out the provisions of this section;

(5) obtain the services of experts and consultants in accordance with the provisions of section 3109 of title 5, United States Code;

(6) accept and utilize the services of voluntary and noncompensated personnel and reimburse them for travel expenses, including per diem, as authorized by section 5703 of title 5, United States Code;

(7) enter into contracts, grants or other arrangements, or modifications thereof to carry out the provisions of this section, and such contracts or modifications thereof may be entered into without performance or other bonds, and without regard to section 3709 of the Revised Statutes, as amended (41 U.S.C. 5), or any other provision of law relating to competitive bidding;

(8) make advance, progress, and other payments which the Director deems necessary under this title without regard to the provisions of section 3648 of the Revised Statutes, as amended (31 U.S.C. 529); and

(9) make other necessary expenditures.

(f) The Director shall submit to the Secretary of Health, Education, and Welfare, to the President, and to the Congress an annual report of the operations of the Institute under this Act, which shall include a detailed statement of all private and public funds received and expended by it, and such recommendations as he deems appropriate.

GRANTS TO THE STATES

SEC. 23. (a) The Secretary is authorized, during the fiscal year ending June 30, 1971, and the two succeeding fiscal years, to make grants to the States which have designated a State agency under section 18 to assist them—

(1) in identifying their needs and responsibilities in the area of occupational safety and health,

(2) in developing State plans under section 18, or

(3) in developing plans for—

(A) establishing systems for the collection of information concerning the nature and frequency of occupational injuries and diseases;

(B) increasing the expertise and enforcement capabilities of their personnel engaged in occupational safety and health programs; or

(C) otherwise improving the administration and enforcement of State occupational safety and health laws, including standards thereunder, consistent with the objectives of this Act.

(b) The Secretary is authorized, during the fiscal year ending June 30, 1971, and the two succeeding fiscal years, to make grants to the States for experimental and demonstration projects consistent with the objectives set forth in subsection (a) of this section.

(c) The Governor of the State shall designate the appropriate State agency for receipt of any grant made by the Secretary under this section.

(d) Any State agency designated by the Governor of the State desiring a grant under this section shall submit an application therefor to the Secretary.

(e) The Secretary shall review the application, and shall, after consultation with the Secretary of Health, Education, and Welfare, approve or reject such application.

(f) The Federal share for each State grant under subsection (a) or (b) of this section may not exceed 90 per centum of the total cost of the application. In the event the Federal share for all States under either such subsection is not the same, the differences among the States shall be established on the basis of objective criteria.

(g) The Secretary is authorized to make grants to the States to assist them in administering and enforcing programs for occupational safety and health contained in State plans approved by the Secretary pursuant to section 18 of this Act. The Federal share for each State grant under this subsection may not exceed 50 per centum of the total cost to the State of such a program. The last sentence of subsection (f) shall be applicable in determining the Federal share under this subsection.

(h) Prior to June 30, 1973, the Secretary shall, after consultation with the Secretary of Health, Education, and Welfare, transmit a report to the President and to the Congress, describing the experience under the grant programs authorized by this section and making any recommendations he may deem appropriate.

STATISTICS

SEC. 24. (a) In order to further the purposes of this Act, the Secretary, in consultation with the Secretary of Health, Education, and Welfare, shall develop and maintain an effective program of collection, compilation, and analysis of occupational safety and health statistics. Such program may cover all employments whether or not subject to any other provisions of this Act but shall not cover employments excluded by section 4 of the Act. The Secretary shall compile accurate statistics on work injuries and illnesses which shall include all disabling, serious, or significant injuries and illnesses, whether or not involving loss of time from work, other than minor injuries requiring only first aid treatment and which do not involve medical treatment, loss of consciousness, restriction of work or motion, or transfer to another job.

(b) To carry out his duties under subsection (a) of this section, the Secretary may—

(1) promote, encourage, or directly engage in programs of studies, information and communication concerning occupational safety and health statistics;

(2) make grants to States or political subdivisions thereof in order to assist them in developing and administering programs dealing with occupational safety and health statistics; and

(3) arrange, through grants or contracts, for the conduct of such research and investigations as give promise of furthering the objectives of this section.

(c) The Federal share for each grant under subsection (b) of this section may be up to 50 per centum of the State's total cost.

(d) The Secretary may, with the consent of any State or political subdivision thereof, accept and use the services, facilities, and employees of the agencies of such State or political subdivision, with or without reimbursement, in order to assist him in carrying out his functions under this section.

(e) On the basis of the records made and kept pursuant to section 8(c) of this Act, employers shall file such reports with the Secretary as he shall prescribe by regulation, as necessary to carry out his functions under this Act.

(f) Agreements between the Department of Labor and States pertaining to the collection of occupational safety and health statistics already in effect on the effective date of this Act shall remain in effect until superseded by grants or contracts made under this Act.

AUDITS

SEC. 25. (a) Each recipient of a grant under this Act shall keep such records as the Secretary or the Secretary of Health, Education, and Welfare shal prescribe, including records which fully disclose the amount and disposition by such recipient of the proceeds of such grant, the total cost of the project or undertaking in connection with which such grant is made or used, and the amount of that portion of the cost of the project or undertaking supplied by other sources, and such other records as will facilitate an effective audit.

(b) The Secretary or the Secretary of Health, Education, and Welfare, and the Comptroller General of the United States, or any of their duly authorized representatives, shall have access for the purpose of audit and examination to any books, documents, papers, and records of the recipients of any grant under this Act that are pertinent to any such grant.

ANNUAL REPORT

SEC. 26. Within one hundred and twenty days following the convening of each regular session of each Congress, the Secretary and the Secretary of Health, Education, and Welfare shall each prepare and submit to the President for transmittal to the Congress a report upon the subject matter of this Act, the progress toward achievement of the purpose of this Act, the needs and requirements in the field of occupational safety and health, and any other relevant information. Such reports shall include information regarding occupational safety and health standards, and criteria for such standards, developed during the preceding year; evaluation of standards and criteria previously developed under this Act, defining areas of emphasis for new criteria and standards; an evaluation of the degree of observance of applicable occupational safety and health standards, and a summary of inspection and enforcement activity undertaken; analysis and evaluation of research activities for which results have been obtained under governmental and nongovernmental sponsorship; an analysis of major occupational diseases; evaluation of available control and measurement technology for hazards for which standards or criteria have been developed during the preceding year; description of cooperative efforts undertaken between Government agencies and other interested parties in the implementation of this Act during the preceding year; a progress report on the development of an adequate supply of trained manpower in the field of occupational safety and health, including estimates of future needs and the efforts being made by Government and others to meet those needs; listing of all toxic substances in industrial usage for which labeling requirements, criteria, or standards have not yet been established; and such recommendations for additional legislation as are deemed necessary to protect the safety and health of the worker and improve the administration of this Act.

NATIONAL COMMISSION ON STATE WORKMEN'S COMPENSATION LAWS

SEC. 27. (a) (1) The Congress hereby finds and declares that—

(A) the vast majority of American workers, and their families, are dependent on workmen's compensation for their basic economic security in the event such workers suffer disabling injury or death in the course of their employment; and that the full protection of American workers from job-related injury or death requires an adequate, prompt, and equitable system of workmen's compensation as well as an effective program of occupational health and safety regulation; and

(B) in recent years serious questions have been raised concerning the fairness and adequacy of present workmen's compensation laws in the light of the growth of the economy, the changing nature of the labor force, increases in medical knowledge, changes in the hazards associated with various types of employment, new technology creating new risks to health and safety, and increases in the general level of wages and the cost of living.

(2) The purpose of this section is to authorize an effective study and objective evaluation of State workmen's compensation laws in order to determine if such laws provide an adequate, prompt, and equitable system of compensation for injury or death arising out of or in the course of employment.

(b) There is hereby established a National Commission on State Workmen's Compensation Laws.

(c) (1) The Workmen's Compensation Commission shall be composed of fifteen members to be appointed by the President from among members of State workmen's compensation boards, representatives of insurance carriers, business, labor, members of the medical profession having experience in industrial medicine or in workmen's compensation cases, educators having special expertise in the field of workmen's compensation, and representatives of the general public. The Secretary, the Secretary of Commerce, and the

Secretary of Health, Education, and Welfare shall be ex officio members of the Workmen's Compensation Commission:

(2) Any vacancy in the Workmen's Compensation Commission shall not affect its powers.

(3) The President shall designate one of the members to serve as Chairman and one to serve as Vice Chairman of the Workmen's Compensation Commission.

(4) Eight members of the Workmen's Compensation Commission shall constitute a quorum.

(d) (1) The Workmen's Compensation Commission shall undertake a comprehensive study and evaluation of State workmen's compensation laws in order to determine if such laws provide an adequate, prompt, and equitable system of compensation. Such study and evaluation shall include, without being limited to, the following subjects: (A) the amount and duration of permanent and temporary disability benefits and the criteria for determining the maximum limitations thereon, (B) the amount and duration of medical benefits and provisions insuring adequate medical care and free choice of physician, (C) the extent of coverage of workers, including exemptions based on numbers or type of employment, (D) standards for determining which injuries or diseases should be deemed compensable, (E) rehabilitation, (F) coverage under second or subsequent injury funds, (G) time limits on filing claims, (H) waiting periods, (I) compulsory or elective coverage, (J) administration, (K) legal expenses, (L) the feasibility and desirability of a uniform system of reporting information concerning job-related injuries and diseases and the operation of workmen's compensation laws, (M) the resolution of conflict of laws, extraterritoriality and similar problems arising from claims with multistate aspects, (N) the extent to which private insurance carriers are excluded from supplying workmen's compensation coverage and the desirability of such exclusionary practices, to the extent they are found to exist, (O) the relationship between workmen's compensation on the one hand, and old-age, disability, and survivors insurance and other types of insurance, public or private, on the other hand, (P) methods of implementing the recommendations of the Commission.

(2) The Workmen's Compensation Commission shall transmit to the President and to the Congress not later than July 31, 1972, a final report containing a detailed statement of the findings and conclusions of the Commission, together with such recommendations as it deems advisable.

(e) (1) The Workmen's Compensation Commission or, on the authorization of the Workmen's Compensation Commission, any subcommittee or members thereof, may, for the purpose of carrying out the provisions of this title, hold such hearings, take such testimony, and sit and act at such times and places as the Workmen's Compensation Commission deems advisable. Any member authorized by the Workmen's Compensation Commission may administer oaths or affirmations to witnesses appearing before the Workmen's Compensation Commission or any subcommittee or members thereof.

(2) Each department, agency, and instrumentality of the executive branch of the Government including independent agencies, is authorized and directed to furnish to the Workmen's Compensation Commission, upon request made by the Chairman or Vice Chairman, such information as the Workmen's Compensation Commission deems necessary to carry out its functions under this section.

(f) Subject to such rules and regulations as may be adopted by the Workmen's Compensation Commission, the Chairman shall have the power to—

(1) appoint and fix the compensation of an executive director, and such additional staff personnel as he deems necessary, without regard to the provisions of title 5, United States Code, governing appointments in the competitive service, and without regard to the provisions of chapter 51 and subchapter III of chapter 53 of such title relating to classification and General Schedule pay rates, but at rates not in excess of the maximum rate for GS–18 of the General Schedule under section 5332 of such title, and

(2) procure temporary and intermittent services to the same extent as is authorized by section 3109 of title 5, United States Code.

(g) The Workmen's Compensation Commission is authorized to enter into contracts with Federal or State agencies, private firms, institutions, and individuals for the conduct of research or surveys, the preparation of reports, and other activities necessary to the discharge of its duties.

(h) Members of the Workmen's Compensation Commission shall receive compensation for each day they are engaged in the performance of their duties as members of the Workmen's Compensation Commission at the daily rate prescribed for GS–18 under section 5332 of title 5, United States Code, and shall be entitled to reimbursement for travel, subsistence, and other necessary expenses incurred by them in the performance of their duties as members of the Workmen's Compensation Commission.

(i) There are hereby authorized to be appropriated such sums as may be necessary to carry out the provisions of this section.

(j) On the ninetieth day after the date of submission of its final report to the President, the Workmen's Compensation Commission shall cease to exist.

ECONOMIC ASSISTANCE TO SMALL BUSINESSES

SEC. 28. (a) Section 7(b) of the Small Business Act, as amended, is amended—

(1) by striking out the period at the end of "paragraph (5)" and inserting in lieu thereof "; and "; and

(2) by adding after paragraph (5) a new paragraph as follows:

"(6) to make such loans (either directly or in cooperation with banks or other lending institutions through agreements to participate on an immediate or deferred basis) as the Administration may determine to be necessary or appropriate to assist any small business concern in effecting additions to or alterations in the equipment, facilities, or methods of operation of such business in order to comply with the applicable standards promulgated pursuant to section 6 of the Occupational Safety and Health Act of 1970 or standards adopted by a State pursuant to a plan approved under section 18 of the Occupational Safety and Health Act of 1970, if the Administration determines that such concern is likely to suffer substantial economic injury without assistance under this paragraph."

(b) The third sentence of section 7(b) of the Small Business Act, as amended, is amended by striking out "or (5)" after "paragraph (3)" and inserting a comma followed by "(5) or (6)".

(c) Section 4(c) (1) of the Small Business Act, as amended, is amended by inserting "7(b) (6)," after "7(b) (5),".

(d) Loans may also be made or guaranteed for the purposes set forth in section 7(b) (6) of the Small Business Act, as amended, pursuant to the provisions of section 202 of the Public Works and Economic Development Act of 1965, as amended.

ADDITIONAL ASSISTANT SECRETARY OF LABOR

SEC. 29. (a) Section 2 of the Act of April 17, 1946 (60 Stat. 91) as amended (29 U.S.C. 553) is amended by—

(1) striking out "four" in the first sentence of such section and inserting in lieu thereof "five"; and

(2) adding at the end thereof the following new sentence, "One of such Assistant Secretaries shall be an Assistant Secretary of Labor for Occupational Safety and Health.".

(b) Paragraph (20) of section 5315 of title 5, United States Code, is amended by striking out "(4)" and inserting in lieu thereof "(5)".

ADDITIONAL POSITIONS

SEC. 30. Section 5108(c) of title 5, United States Code, is amended by—

(1) striking out the word "and" at the end of paragraph (8);

(2) striking out the period at the end of paragraph (9) and inserting in lieu thereof a semicolon and the word "and"; and

(3) by adding immediately after paragraph (9) the following new paragraph:

"(10) (A) the Secretary of Labor, subject to the standards and procedures prescribed by this chapter, may place an additional twenty-five positions in the Department of Labor in GS–16, 17, and 18 for the purposes of carrying out his responsibilities under the Occupational Safety and Health Act of 1970;

"(B) the Occupational Safety and Health Review Commission, subject to the standards and procedures prescribed by this chapter, may place ten positions in GS–16, 17, and 18 in carrying out its functions under the Occupational Safety and Health Act of 1970."

EMERGENCY LOCATOR BEACONS

SEC. 31. Section 601 of the Federal Aviation Act of 1958 is amended by inserting at the end thereof a new subsection as follows:

"EMERGENCY LOCATOR BEACONS

"(d) (1) Except with respect to aircraft described in paragraph (2) of this subsection, minimum standards pursuant to this section shall include a requirement that emergency locator beacons shall be installed—

"(A) on any fixed-wing, powered aircraft for use in air commerce the manufacture of which is completed, ⊙ which is imported into the United States, after one year following the date of enactment of this subsection; and

"(B) on any fixed-wing, powered aircraft used in air commerce after three years following such date.

"(2) The provisions of this subsection shall not apply to jet-powered aircraft; aircraft used in air transportation (other than air taxis and charter aircraft); military aircraft; aircraft used solely for training purposes not involving flights more than twenty miles from its base; and aircraft used for the aerial application of chemicals."

SEPARABILITY

SEC. 32. If any provision of this Act, or the application of such provision to any person or circumstance, shall be held invalid, the remainder of this Act, or the application of such provision to persons or circumstances other than those as to which it is held invalid, shall not be affected thereby.

APPROPRIATIONS

SEC. 33. There are authorized to be appropriated to carry out this Act for each fiscal year such sums as the Congress shall deem necessary.

EFFECTIVE DATE

SEC. 34. This Act shall take effect one hundred and twenty days after the date of its enactment.

Approved December 29, 1970.

Notes

Chapter 1

1. U. S. House of Representatives Report No. 1291 on H. R. 16785, 91st Congress, 2nd Session (July 9, 1970), p. 14.
2. Herman and Ann Somers, *Workmen's Compensation: Prevention, Insurance and Rehabilitation of Occupational Disability* (New York, 1954), p. 6.
3. Hearings on S. 2193, S. 2788 before the Subcommittee on Labor of the U. S. Senate Committee on Labor and Public Welfare, 91st Congress, 1st and 2nd Sessions, part 1 (1969–70), p. 647 (hereinafter cited as 1970 Senate Hearings).
4. *Occupational Disease: The Silent Enemy*, U. S. Department of Health, Education and Welfare (HEW), Public Health Service (PHS) publication (Washington, D. C., 1968), p. 1.
5. Testimony of William Stewart, M. D., U. S. Surgeon General, on coal mine health and safety before the Subcommittee on Labor of the Senate Committee on Labor and Public Welfare, 91st Congress, 1st Session, part 2 (1969), p. 720 (hereinafter cited as Senate Coal Mine Health and Safety Hearings).
6. *Ibid.*, p. 674.
7. *President's Report on Occupational Safety and Health*, Commerce Clearing House ed. (May 22, 1972), p. 111.
8. C. Dean McClure, M. D., "An Occupational Health Survey of an Urban Area," Table III, paper reprinted in Occupational Safety and Health Hearings before the Select Subcommittee on Labor of the House Committee on Education and Labor, 91st Congress, 1st Session, part 2 (1969), p. 1230 (hereinafter cited as 1969 House Hearings).
9. *Ibid.*, p. 1228.
10. Testimony of William Stewart, M. D., U. S. Surgeon General, Hearings on H.R. 14816, Occupational Safety and Health, before the Select Subcommittee on Labor of the House Committee on Education and Labor, 90th Congress, 2nd Session (1968), pp. 106–107 (hereinafter cited as 1968 House Hearings).
11. *Occupational Disease, supra* note 4, p. 6.
12. Brit Hume, *Death and the Mines* (New York, 1971), pp. 112–113.
13. Neal Herrick and Robert Quinn, "The Working Conditions Survey as a Source of Social Indicators," *Monthly Labor Review* (April, 1971).
14. *Hazards in the Industrial Environment*, District 8 Council, Oil, Chemical and Atomic Workers International Union (OCAW) Conference,

Kenilworth, N.J. (March 29, 1969), (reprinted in 1969 House Hearings, p. 1267).

15. *Ibid.*, p. 1265.
16. *Ibid.*, pp. 1292–1293.

Chapter 2

1. 1970 Senate Hearings, part 1, p. 855.
2. Testimony of William H. Stewart, M. D., U. S. Surgeon General, 1968 House Hearings, p. 107.
3. *Occupational Characteristics of Disabled Workers, by Disabling Conditions: HEW Disability Insurance Benefit Awards Made in 1959–1962 to Men under Age 65,* HEW, PHS Publication No. 1531, pp. 1, 160.
4. *Monthly Labor Review* (December, 1968), p. 87.
5. Donald Hunter, *The Diseases of Occupations,* 4th ed. (London, 1969), p. 955.
6. Reprinted in Harvey Swados, ed., *Years of Conscience: The Muckrakers* (New York, 1962), p. 180.
7. 1969 House Hearings, part 1, p. 36.
8. *Supra* note 5, ch. 1.
9. *Ibid.*, part 2, p. 730.
10. 1970 Senate Hearings, part 1, pp. 983–984.
11. Eugenija Zuskin, M. D., *et al.*, "Byssinosis in Carding and Spinning Workers: Prevalence in the Cotton Textile Industry," *Archives of Environmental Health* (1969), (reprinted in 1970 Senate Hearings, part 1, p. 990).
12. Arend Boyhuys, "Byssinosis in Textile Workers," paper presented to Division of Environmental Sciences, Yale University School of Medicine (January 5, 1966), (reprinted in 1970 Senate Hearings, part 2, p. 1568).
13. *Ibid.*, p. 1575.
14. *HEW Activities in Byssinosis,* HEW Environmental Health Service, Bureau of Occupational Safety and Health publication (Washington, D. C., 1970).
15. Editorial in *America's Textile Reporter* (July 10, 1969), (reprinted in 1970 Senate Hearings, part 1, p. 999).
16. Bryan Gandevia, M. D., "Transactions of the National Conference on Cotton Dust and Health," paper presented to Continuing Education and Field Service, School of Public Health, University of North Carolina (May 2, 1970), p. 63.
17. 1970 Senate Hearings, part 2, p. 1074.
18. Paul Brodeur, *Asbestos and Enzymes* (New York, 1972), p. 14.
19. Irving J. Selikoff, M. D., E. Cuyler Hammond, Sc. D., and Jacob Churg, M. D., "Mortality Experience of Asbestos Insulation Workers, 1912–1971," paper presented at the 4th International Conference on Pneumoconiosis, Bucharest (September 29, 1971).
20. *Ibid.*
21. 1970 Senate Hearings, part 2, p. 1075.
22. Brodeur, *supra* note 18, p. 21.
23. *Ibid.*, p. 25.
24. *Ibid.*, p. 39.

25. Selikoff, *et al., supra* note 19.
26. *Ibid.*, p. 4.
27. *Loc. cit.*
28. *Loc. cit.*
29. Telephone communication with Study Group member (September 10, 1971).
30. Brodeur, *supra* note 18, p. 40.
31. *Loc. cit.*
32. Clark Whelton, "Asbestos: Four More Months," *Village Voice* (October 21, 1971), p. 20.
33. Hunter, *supra* note 5, p. 819.
34. J. William Lloyd, "Long Term Mortality Study of Steelworkers, Part V: Respiratory Cancer in Coke Plant Workers," *Journal of Occupational Medicine* (February, 1971), p. 67.
35. *Loc. cit.*
36. Leo Goodman, "Radiation Hazard in Modern Industry," paper presented to the John Fogarty Memorial under the auspices of the American Public Health Association Medical Care Section and the District of Columbia Public Health Association (April 26, 1967).
37. Frank Lundin, Jr., *et al.*, "Mortality of Uranium Miners in Relation to Radiation Exposure, Hard Rock Mining and Cigarette Smoking, 1950 through September 1967," p. 21.
38. Hunter, *supra* note 5, p. 817.
39. *Ibid.*, p. 818.
40. "Bladder Cancer and Occupations," *British Medical Journal* (March 6, 1971).
41. *Petroleum Facts and Figures*, American Petroleum Institute (1967 ed.), p. 222.
42. *Lancet* (November 7, 1970), p. 967.
43. H. G. Parkes, "Cancer and Exposure to Mineral Oils," *Industrial Medicine and Surgery* (February, 1970), p. 80.
44. *Lancet, supra* note 42, p. 967.
45. "The Case of Takoko Nakamura," *Parade* (April 18, 1971), p. 9.
46. Blejer, M. D., "Death Due to Cadmium Oxide Fumes," *Industrial Medicine and Surgery* (May, 1966), pp. 363–364.
47. Marion Edey, UAW *Washington Report* (July 19, 1971).
48. UAW *Washington Report* (June 7, 1971).
49. *New York Times* (June 3, 1971), p. 70.
50. Copyright 1971, by Newsday. Distributed by Los Angeles Times Syndicate. (Reprinted in UAW *Washington Report*, June 14, 1971).
51. Testimony of George Burke, staff representative, United Steelworkers of America, 1969 House Hearings, part 2, p. 1002.
52. *Ibid.*, pp. 1002–1003.
53. 1969 House Hearings, p. 1292.
54. *New York Times* (June 3, 1971), p. 70.
55. *Hazards in the Industrial Environment*, District 8 Council, OCAW Conference (June 14, 1969), p. 54.
56. *Hazards in the Industrial Environment,* District 5 Council, OCAW Conference (November 1, 1969), p. 12.
57. 1969 House Hearings, pp. 1268–1271.
58. *Preventing Accidental Ozone Poisoning in Workers*, HEW, PHS publication No. 1526.

59. H. E. Stokinger, Ph. D., "Ozone Toxicology," *Archives of Environmental Health* (May, 1965).

60. T. Higgins, M. D., "Effects of Sulfur Oxides and Particulates on Health," *Archives of Environmental Health* (May, 1971).

61. *Ibid.*, p. 584.

62. *Ibid.*, p. 589.

63. "In-plant Air is Danger to Health," UAW *Washington Report* (April 20, 1970).

64. *Hazards in the Industrial Environment, supra* note 56, p. 8.

65. John Esposito, *Vanishing Air* (New York, 1970), p. 15.

66. *Hazards in the Industrial Environment, supra* note 56, p. 8.

67. *Ibid.*, p. 9.

68. Quoted in statement of Anthony Mazzocchi, citizenship-legislative director, OCAW, 1969 House Hearings, part 2, p. 1182.

69. Hearings on S.2864, Occupational Safety and Health Act of 1968, before the Subcommittee on Labor of the Senate Committee on Labor and Public Welfare, 90th Congress, 2nd Session (1968), p. 108 (hereinafter cited as 1968 Senate Hearings).

70. May Mayers, *et al.*, "Benzene (benzol) Poisoning in the Rotogravure Printing in New York City," *Journal of Industrial Hygiene and Toxicology* (October, 1939), pp. 395–420.

71. Interview with Study Group member (September, 1971).

72. Telephone communication from the American Petroleum Institute (September 27, 1971).

73. Hunter, *supra* note 5, p. 515.

74. Horace W. Gerarde, "The Aromatic Hydrocarbons," in Frank A. Patty, ed., *Industrial Hygiene and Toxicology* (New York, 1963), p. 1226.

75. *Hazards in the Industrial Environment*, District 4 Council, OCAW Conference (February, 1970), p. 29.

76. Alfred Jasser, "Alert Your Plant Workers to Effects of Dangerous Chemicals," International Printing Pressmen and Assistants Union of North America, AFL-CIO.

77. Don Irish, "Alphatic Hologenated Hydrocarbons," in Patty, *supra* note 74, p. 1265.

78. *Federal Register*, Vol. 35 (August 19, 1970), p. 13198.

79. Interview with Donald Whorton, M. D., Health Research Group, Washington, D. C., (1971).

80. V. K. Rowe and M. A. Wolf, "Ketones," in Patty, *supra* note 74, p. 1726.

81. G. H. Hine and V. K. Rowe, "Epoxy Compounds," in Patty, *supra* note 74, p. 1597.

82. John M. Peters *et al.*, "Respiratory Impairment in Workers Exposed to 'Safe' Levels of Toluene Diisocyanate (TDI)," *Archives of Environmental Health* (March, 1970).

83. 1969 House Hearings, part 2, p. 1356.

84. Robert Fellmeth, *Power and Land in California: Ralph Nader Task Force Report* (Washington, D. C., 1971), II:184.

85. Steve Wodka, "Pesticides Since Silent Spring," in Garrett de Bell, ed., *Environmental Handbook* (New York, 1970), p. 85.

86. Fellmeth, II:186.

87. *Ibid.*, II:188.

88. *Science* (January 8, 1971), p. 43.
89. Quoted in Harrison Wellford, *Sowing the Wind* (New York, 1972), p. 197.
90. *Loc. cit.*
91. *Science* (January 8, 1971), p. 46.
92. Thomas Whiteside, *Defoliation* (New York, 1970), pp. 20–22.
93. *Science* (April 24, 1970), p. 453.
94. Wellford, *supra* note 89, p. 201.
95. *Science* (April 24, 1970), p. 453.
96. Wellford, *supra* note 89, p. 201.
97. "Threshold Limit Values for 1969," American Conference of Governmental Industrial Hygienists (reprinted in 1970 Senate Hearings, part 2, p. 1249).
98. "Issue Study on Noise Control," HEW, PHS publication (1969), reprinted in 1970 Senate Hearings, part 2, p. 1640.
99. Joseph R. Anticaglia, M. D., "National Noise Study," paper presented to the First International Conference on Occupational Safety and Health, United Steelworkers of America, Cincinnati, Ohio (November, 1969), p. 203.

Chapter 3

1. E. H. Downey, *History of Work Accident Indemnity in Iowa* (1912), quoted in E. E. Cummings, *The Labor Problem in the United States,* 2nd ed. (New York, 1935), pp. 73–74.
2. John Commons, *Trade Unionism and Labor Problems* (Boston, 1905), p. 437.
3. Upton Sinclair, *The Jungle* (Airmont, New York, 1965), p. 115.
4. William Hard, "The Law of the Killed and Wounded," *Everybody's Magazine* (September, 1908), p. 361.
5. David Brody, *Steelworkers in America: The Nonunion Era* (New York, 1969), p. 34.
6. Crystal Eastman, *Work-Accidents and the Law* (Russell Sage Foundation, 1910), p. 95.
7. Brody, *supra* note 5, p. 92.
8. William Hard, "Making Steel and Killing Men," *Everybody's Magazine* (November, 1907), p. 579.
9. Arno Dosch, "Just Wops," *Everybody's Magazine* (November, 1911), p. 579.
10. Alice Hamilton, *Exploring the Dangerous Trades* (Boston, 1943), pp. 151–152.
11. Brody, *supra* note 5, p. 100.
12. Eastman, *supra* note 6.
13. John Commons and John Andrews, *Principles of Labor Legislation,* 4th ed. rev. (New York, 1936), p. 164.
14. Donald Hunter, *The Diseases of Occupations,* 4th ed. (London, 1969), p. 378.
15. Walter F. Dodd, *Administration of Workmen's Compensation* (London, 1936), p. 698.
16. Lew Palmer, "History of the Safety Movement," *The Annals* (January, 1926), p. 19.

17. Robert Dunn, *Labor and Automobiles* (New York, 1929), p. 147.
18. Cummings, *supra* note 1, p. 585.
19. *Loc. cit.*
20. Hearings on Appropriations for the Department of HEW, Senate Committee on Appropriations (1961), p. 709.
21. "An Investigation Relating to Health Conditions of Workers Employed in the Construction and Maintenance of Public Utilities," Hearings before Special Subcommittee of the House Committee on Labor (1936). The account of the Gauley Bridge disaster is taken from these hearings (hereinafter cited as Gauley Bridge Hearings).
22. *Ibid.*, p. 9.
23. Vito Marcantonio, "Dusty Death," *New Republic* (March 5, 1936).
24. Gauley Bridge Hearings, p. 112.
25. *Congressional Record*, 74th Congress (March 1, 1936), H4752.
26. Gauley Bridge Hearings, p. 129.
27. *Ibid.*, p. 203.
28. Hearings before Special Subcommittee of the Senate Committee on Education and Labor (May 13, 1940).
29. Hearings to Establish Safe and Healthful Working Conditions in Industry before Special Subcommittee of the House Committee on Labor (1943).
30. Hearings on Industrial Safety before the Senate Committee on Labor and Public Safety (March 24, 1952).
31. Hearings on Occupational Safety before the General Subcommittee on Labor of the House Committee on Education and Labor (April 17, 1962).
32. *State Workmen's Compensation Laws: A Comparison of Major Provisions with Recommended Standards,* U. S. Department of Labor, BLS, Bulletin 212 (rev. 1961).
33. *Ibid.* (rev. 1967).
34. *Annual Report of Compensable Work Injuries*, Illinois Industrial Commission (1968), II:17.
35. Earl Cheit, *Injury and Recovery in the Course of Employment* (New York, 1961), pp. 62–68, 106–109.
36. Monroe Berkowitz and John Burton, "Income Maintenance Objective of Workmen's Compensation," *Industrial and Labor Relations Review* (October, 1970), p. 17.
37. *Ibid.*, p. 23.
38. *The Workmen's Disability Income System: Recommendations for Federal Action*, Confidential Report of Task Force to the President's Council of Economic Advisors (August, 1968), p. 108.
39. For Larson's own description of the experience, see Arthur Larson, "Compensation Reform in the United States," *Occupational Disability and Public Policy* (New York, 1963), pp. 26–31.
40. Earl Cheit, "The Disability Benefit Complex," *Rutgers Law Journal* Vol. 18 (1964), p. 565.
41. *Richardson v. Belcher*, 92 S. Ct. 254 (U. S. 1971).

Chapter 4

1. *Estimated Employee Coverage Under State Safety Rulemaking Authority*, U. S. Department of Labor, BLS publication (July, 1969).

2. *Ibid.*
3. *Accident Facts*, National Safety Council (1970), p. 23.
4. "Table I: Comparison of Number of Safety Inspectors with Number of Fish and Game Wardens Employed by Specified States, 1968," Appendix to testimony by George Meany, president, AFL-CIO, 1968 House Hearings, p. 711.
5. Interview with Mr. John Proctor, U. S. Department of Labor, BLS, Office of Standards (August 10, 1970).
6. *State Budget Allocations and Staffing for Occupational Safety*, U. S. Department of Labor, BLS, Division of Programming and Research (June, 1969).
7. *Annual Report* (fiscal year 1967–1968), Indiana Division of Labor, p. 1.
8. Coroner's verdict, State of Indiana (March 17, 1970), (reprinted in 1970 Senate Hearings, part 2, pp. 1587–1589).
9. *State Enforcement Provisions*, U. S. Department of Labor, BLS, Office of Occupational Safety (1969).
10. *Loc. cit.*
11. Letter to Senator Harrison Williams (D.-N.J.), chairman, Senate Subcommittee on Labor, from T. R. Johnson, executive director, Ohio Manufacturers' Association (September 26, 1969), (reprinted in 1970 Senate Hearings, part 2, p. 1152).
12. *Supplement to 1968 Annual Report* (1969), New York State Department of Labor Statistics on Operations, pp. 72, 77.
13. Jack Newfield, "Adam and the Hardhats: Facts Make a Dent," *Village Voice* (October 15, 1970), p. 1.
14. *Annual Report* (1968), Massachusetts Division of Industrial Safety, Department of Labor and Industry, p. 19.
15. Survey for the Safety Committee of the International Association of Industrial Accident Boards and Commissions (IAIABC), tabulated by the U. S. Department of Labor, BLS (July, 1970).
16. Interview with Thomas Seymour, U. S. Department of Labor, BLS (June 29, 1970).
17. IAIABC Survey (1970), *supra* note 15.
18. Letter from George Hagglund, School for Workers, University of Wisconsin, to the Study Group (December 1, 1970).
19. Eugene Meyer, "Office Held Lax on D. C. Job Safety," *Washington Post* (February 2, 1971), p. B-1.
20. 1970 Senate Hearings, part 1, p. 809.
21. *Ibid.*, p. 904.
22. W. Dean Keefer, "The New Deal and Safety," *IAIABC Convention Proceedings* (1933), p. 178.
23. *Monitor*, Vol. 42, No. 1, p. 14 (publication of the Ohio Industrial Commission, on file at the U. S. Department of Labor, BLS, Washington, D. C.).
24. Testimony of Jacob Clayman, administrative director, Industrial Union Department, AFL-CIO, before 1970 Senate Hearings, part 1, p. 422.
25. Letter from Holland Krise, chairman of the Ohio Industrial Commission, to the Study Group (October 2, 1970).
26. *Focus* (January, 1969), p. 5 (publication of the Ohio AFL-CIO on file at the U. S. Department of Labor, BLS, Washington, D. C.).
27. 1969 House Hearings, part 2, p. 866.

28. *IAIABC Convention Proceedings* (1969), p. 153.
29. Victoria Trasko, *Status of Occupational Health Programs in State and Local Government*, HEW, Bureau of Occupational Safety and Health (BOSH) (1969).
30. 1970 Senate Hearings, part 1, p. 137.
31. *Issue Study on Occupational Health*, HEW, BOSH (reprinted in 1970 Senate Hearings, part 2, p. 1702).
32. *State and Local Personnel Resources Available for Essential Industrial Hygiene Services*, HEW, BOSH (1968).
33. Trasko, *supra* note 29, p. 2.
34. *Loc. cit.*
35. "Threshold Limit Values of Airborne Contaminants for 1969," American Conference of Governmental Industrial Hygienists (reprinted in 1970 Senate Hearings, part 2, p. 1249).
36. Andrew D. Hosey and Lorice Ede, J. D., *A Review of State Occupational Health Legislation*, HEW, BOSH (1969), (reprinted in 1970 Senate Hearings, part 1, pp. 126–127).
37. *Florida Occupational Health Survey*, HEW, BOSH (1966), p. 2.
38. *Chicago Occupational Health Survey*, HEW, BOSH (1968).
39. *Region V Report*, HEW, BOSH (1969), p. 2.
40. 19 Conn. Gen. Stat. Ann. 1 (1958) and 71 Conn. Gen. Stat. Ann. 9–11 (1958).
41. California Labor Code 6314 (West's 1955); California Labor Code 3760 and 6407 (Supp. 1969).
42. Victoria Trasko, *Occupational Health and Safety Legislation*, HEW, PHS (1970).
43. 1968 Senate Hearings, p. 88.

Chapter 5

1. 1970 Senate Hearings, part 1, pp. 138–139.
2. 1969 House Hearings, part 1, p. 746.
3. Reprinted in 1970 Senate Hearings, part 2, p. 1727.
4. *Federal Register*, Vol. 33, No. 247 (December 20, 1968), p. 19051.
5. 1969 House Hearings, Part 1, p. 747.
6. *American Journal of Public Health*, Vol. 61 (August, 1971), pp. 1583–1585.
7. Interview with Dr. Marcus Key, HEW, BOSH (September 2, 1969).
8. "Danger: Men at Work," 1969 Occupational Safety and Health Conference, Industrial Union Department, AFL-CIO, p. 28.
9. 1968 Senate Hearings, p. 65.
10. *Congressional Record* (April 29, 1952), S4557.
11. Interview with Eugene Newman, chief, Contract Safety Division, U. S. Department of Labor, BLS (June 25, 1969).
12. Brit Hume, *Death and the Mines* (New York, 1971), p. 190.
13. *Wecksler* v. *Shultz*, 324 F. Supp. 1085 (D.D.C. 1971).
14. These figures were compiled with the assistance of James Miller, Deputy Associate Solicitor for Litigation, and from the Department of Labor Bulletin, *Safety Engineering and Program Services* (1969).
15. Interview with James Miller (Summer, 1969).

16. Interview with Louis Jacobs, U. S. Department of Labor, New York Regional Office (July 22, 1971).
17. "Walsh-Healey Public Contracts Act Cases Instituted under Safety and Health Regulations During Fiscal Years 1965, 1966, 1967, 1968, 1969," compiled by the U. S. Department of Labor (July 7, 1969), for the Study Group.
18. *Zero In Program Planning Guide*, National Safety Council.
19. Letter to Representative Daniels from George Guenther, enclosing reply to Eugene Newman, Office of Compliance, from John Galuardi, regional administrator, General Services Administration (August 9, 1971).
20. Congressional Delegate Walter E. Fauntroy, news release (January 5, 1972).
21. Report of R. A. Kosser, General Services Administration (October 18, 1963).
22. David Boldt, "GSA Transfers Pair from Heat Tunnel Job," *Washington Post* (January 15, 1972), p. A-20.
23. Austin Henschel, M. D., National Institute for Occupational Safety and Health, in telephone conversation with Robert Vaughn, Public Interest Research Group (May 17, 1971).
24. "Memorandum for the Heads of Executive Departments and Agencies," *Federal Register Document 65-2096* (February 16, 1965).
25. *Mission Safety-70*, Report to the President by the Secretary of Labor (July, 1969), p. 5.
26. "Safety Policy for the Federal Service," reprinted in *Handbook: Federal Safety Council,* U. S. Department of Labor (1963).
27. Interview with Fred Bishoff, director of safety, U. S. Post Office Department (Summer, 1970).
28. *Method of Recording and Measuring Work Injury Experience*, United States of America Standards Institute (December 27, 1967), p. 19.
29. *Ibid.*, p. 23.
30. Letter to John Carwell, chief, Space and Property Management Division, U. S. Department of State from Allen Heins, senior assistant sanitary engineer (May 20, 1969).
31. U. S. Civil Service Commission, Federal Personnel Manual Letter No. 532–17 (August 5, 1970), Attachment 1.

Chapter 6

1. William Green, "The Need for Safety from the Worker's Point of View," *Annals* (January, 1926), p. 5.
2. 1968 House Hearings, p. 244.
3. *Safety Clauses in Union Contracts in New York State, 1956,* Department of Labor, State of New York.
4. *Recommended Safety and Health Program*, United Steelworkers of America, p. 17 (copy on file with the authors).
5. Henry Still, *In Quest of Quiet* (Harrisburg, Pennsylvania, 1970), p. 149.
6. *Op. cit. supra* note 3.
7. 1968 House Hearings, p. 142.
8. Agreement between Kennecott Refining Corp. and the United Steel-

workers of America, Local 5977, Anne Arundel County, Maryland (January 21, 1965).

9. *Hazards in the Industrial Environment*, District 8 Council, Oil, Chemical and Atomic Workers International Union (OCAW) Conference, Kenilworth, New Jersey (March 29, 1969), (reprinted in 1969 House Hearings, p. 1267).

10. Interview with Anthony Mazzocchi, legislative director, OCAW (July 9, 1970).

11. *Hazards in the Industrial Environment, supra* note 9, p. 72. 1969 House Hearings, p. 1306.

12. 1968 House Hearings, p. 390.

13. *Ibid.*, p. 158.

14. *Ibid.*, p. 536.

15. First National Safety Conference, United Steelworkers of America, Chicago, Illinois (November 10, 1969), pp. 50–51.

16. Susan Cottingham, "Tactics Study," paper presented to the Alliance for Labor Action, Washington, D. C. (1971).

17. Anthony Mazzocchi, *supra* note 10.

18. Letter to Ralph Nader (October 29, 1970).

19. Testimony of Lorin Kerr, M. D., Senate Coal Mine Health and Safety Hearings, p. 676.

20. 1968 House Hearings, p. 737.

21. *Ibid.*

22. Letter to the Study Group from Hawey A. Wells, M. D. (February 16, 1972).

23. Joint Occupational Health Program Memorandum of Agreement, United Rubber Workers and B. F. Goodrich Co. (June 13, 1970).

24. OCAW Bargaining Clauses, Appendix B.

25. "Danger: Men at Work," 1969 Occupational Safety and Health Conference, Industrial Union Department, AFL-CIO, p. 28.

26. Still, *supra* note 5, p. 149.

27. Letter to Ralph Nader (January 12, 1970).

28. Letter to Ralph Nader (October 29, 1970).

29. Interview with Anthony Mazzocchi (November 18, 1971).

Chapter 7

1. *Congressional Record* (January 23, 1968), S589, H688.

2. Remarks to President's Conference on Occupational Safety (June 23, 1964), in *Public Papers of the Presidents, Lyndon B. Johnson*, I:808.

3. *Loc. cit.*

4. Remarks at Oath Taking of John W. Gardner, Secretary of HEW (August 18, 1965), in *Public Papers*, II:893.

5. Remarks to members of the International Press Association (May 23, 1966), in *Public Papers*, I:539.

6. Remarks at signing of the Federal Metallic and Non-Metallic Mine Safety Act (September 16, 1966), in *Public Papers*, II:1030.

7. Remarks of Secretary of Labor Willard Wirtz, Third Anniversary of Mission Safety-70 (February 16, 1968).

8. 1968 House Hearings, p. 250.

9. *Ibid.*, p. 256.
10. *Ibid.*, p. 34.
11. 1968 Senate Hearings, p. 636.
12. *Ibid.*, p. 637.
13. *Nation's Business*, Vol. 56, No. 4 (April, 1968), p. 37.
14. 1968 House Hearings, p. 24.
15. Remarks of Senator Jacob Javits (R.-N.Y.), *Congressional Record* (August 6, 1969), S9297.
16. John A. Donnelly, "How Do You Motivate for Safety?" *Environmental Control and Safety Management* (December, 1970), p. 23.
17. *Wall Street Journal* (February 17, 1972).
18. Donnelly, *supra* note 16, p. 23.
19. H.R. 13373, Section 17(a)(3) (August 6, 1969).
20. 1968 House Hearings, p. 234.
21. *Ibid.*, p. 187.
22. 1970 Senate Hearings, part 1, p. 260.
23. House Report No. 90–1720 on H.R. 17748, 90th Congress, 2nd Session (July 16, 1968).
24. Alice Hamilton, *Exploring the Dangerous Trades* (Boston, 1943), pp. 153–154.
25. Crystal Eastman, *Work-Accidents and the Law* (Russell Sage Foundation, 1910), p. 84.
26. Carroll R. Daugherty, *Labor Problems in American Industry*, 5th ed. (Boston, 1941), p. 106.
27. J. Michael Harrison, "Industrial Health and Safety: The Need for Extended Federal Regulation," *Prospectus*, Vol. 3 (December, 1969), p. 180.
28. Letter from L. L. Mollere, chief, Inspection Services, Industrial Safety and Building Division, Department of Industry, Labor and Human Relations, State of Wisconsin, to Dr. George Hagglund (November 10, 1969); letter from Mr. Mollere to Prof. Joseph A. Page (July 20, 1970).
29. Letter from Mr. Mollere to Prof. Page (May 19, 1971).
30. 1968 Senate Hearings, p. 622.
31. "The Rising Tide of Drug Abuse in Industry," *Occupational Hazards* (December, 1971), p. 29; "Drugs in Industry, A Growing Problem," *The Attack* (New York State Narcotic Addiction Control Commission), Vol. 5, No. 2, p. 3.
32. 1968 Senate Hearings, p. 517.
33. 1969 House Hearings, part 2, p. 866.
34. Testimony of A. C. Blackman, secretary and managing director, American Society of Safety Engineers, 1968 Senate Hearings, p. 208.
35. *Loc. cit.*
36. *Status Report—Education Opportunities in Safety—July 1969*, American Society of Saftey Engineers (reprinted in 1969 House Hearings, part 1, p. 526).
37. 1970 House Hearings, part 1, p. 777.
38. 1968 House Hearings, p. 225.
39. Interview with George Wilkins, M. D. (July 22, 1969).
40. 1970 Senate Hearings, part 1, pp. 734–735.
41. *Membership Information Bulletin*, Industrial Hygiene Foundations of America, Inc., Pittsburgh, Pennsylvania, p. 17.

42. *Ibid.*, p. 14.
43. Industrial Hygiene Foundation Bulletin No. 42, 33rd Annual Meeting, Pittsburgh, Pennsylvania (1968).
44. 1970 Senate Hearings, part 1, p. 368.
45. Letter to Robert Harris, counsel, Senate Subcommittee on Labor, from Andrew Kalmykow, counsel, American Insurance Association (July 11, 1968), (reprinted in 1968 Senate Hearings, p. 814).
46. Testimony of Frederick H. Deeg, safety engineer, American Mutual Insurance Alliance, 1968 House Hearings, p. 438.
47. Interview with Robert Heitzman, assistant general manager, National Council of Compensation Insurers (Summer, 1970).
48. Testimony of Andrew Kalmykow, 1968 House Hearings, p. 428.
49. 1968 Senate Hearings, p. 814.
50. *Threshold Limit Values of Airborne Contaminants for 1969,* American Conference of Governmental Industrial Hygienists, Preface (reprinted in 1970 Senate Hearings, part 2, p. 1250).
51. 1970 Senate Hearings, part 1, p. 628.
52. See Opala, "The Anatomy of Private Standards-Making Process," *Oklahoma Law Review*, vol. 22 (1969), pp. 45, 50.
53. "Walsh-Healey Public Contracts Act, McNamara-O'Hara Services Contract Act," U. S. American Standards Institute (USASI), Selected Standards.
54. *Final Report of the National Commission on Product Safety* (June, 1970), p. 53.
55. *Loc. cit.*
56. *Loc. cit.*
57. *Loc. cit.*
58. Interview, American National Standards Institute (ANSI) (Summer, 1969).
59. *USASI Constitution and Bylaws*, B 6.3 (1966), p. 16.
60. *Ibid.*, Art. 3, Basic Principles, p. 5.
61. National Commission on Product Safety, *supra* note 54, p. 52.
62. *Status of Safety Standards*, U. S. Department of Labor (1968), p. 9.
63. "Technical and Financial Growth First 6 Months," report of managing director, ANSI (August 6, 1969), p. 1.
64. *Accident Facts*, National Safety Council (1971 ed.), pp. 3, 5, 23–24.
65. *Injury Rate by Industry, 1969*, U. S. Department of Labor, BLS Report 360, p. 1.
66. "Method of Recording and Measuring Injury Experience," USASI (December 27, 1967), p. 8.
67. *Proceedings of the Sixteenth Annual Meeting of the International Association of Industrial Accident Boards and Commissions,* U. S. Department of Labor, BLS Bulletin 511, pp. 161–164.
68. 1970 Senate Hearings, part 1, pp. 695, 697.
69. Testimony of Anthony Semararo, Fairless Works, Fairless Hills, Pennsylvania, 1970 Senate Hearings, part 1, p. 814.
70. 1968 House Hearings, p. 528.
71. Letter from Milton Durham, corporate director for safety, Martin-Marietta Corp., to Representative Phillip Burton (D.-Calif.) (August 15, 1969); letter from R. F. Wiley, manager, Bethlehem Steel Corp., to Representative Burton (August 20, 1969).
72. Quoted in Charles Pearce, "Quality of Statistics on Work Injury

Rates," paper presented at Interstate Conference on Labor Statistics, Knoxville, Tennessee (July 9, 1959), pp. 13, 14.

73. 1970 Senate Hearings, part 1, p. 205.
74. *Introduction to Bureau of Labor Statistics: Handbook of Methods of Surveys and Studies*, U. S. Department of Labor, BLS Bulletin 1458, p. 1.
75. 1970 Senate Hearings, part 1, p. 630.
76. Jerome Gordon *et al.*, "An Evaluation of the National Industrial Safety Statistics Program," report submitted to the Office of Planning, Wage and Labor Standards Administration, U. S. Department of Labor (June 30, 1970).
77. *Loc. cit.*
78. 1970 Senate Hearings, part 1, p. 855.
79. Senate Coal Mine Health and Safety Hearings, p. 722.
80. 1970 Senate Hearings, part 1, p. 609.
81. 1969 House Hearings, part 1, p. 769.

Chapter 8

1. 1968 Senate Hearings, pp. 61, 62.
2. 1970 Senate Hearings, part 1, p. 81.
3. House Report No. 91–1291, Occupational Safety and Health Act, 91st Congress, 2nd Session (July 19, 1970), p. 19.
4. Morton Mintz, "Lines Harden on Job Safety Legislation," *Washington Post* (October 2, 1970), p. 19.
5. See James M. Estep, "The Drafting of the Daniels Bill," paper prepared for the course on legislation at the Georgetown University Law Center (May 22, 1972).
6. S. 2788, Section 7(a)(1), 91st Congress, 1st Session (August 6, 1969).
7. Remarks by Representative Dominick Daniels, *Congressional Record* (November 23, 1970), H10624.
8. P.L. 91–596, Section 8(c)(2).
9. *Ibid.*, Section 2(b).
10. *Ibid.*, Section 6(b)(5).
11. *Ibid.*, Section 13(a).
12. *Ibid.*, Section 17(k).
13. *Ibid.*, Section 20(a)(7).
14. *Ibid.*, Section 6(a).
15. *Ibid.*, Section 20(a)(6).
16. *Ibid.*, Section 8(c)(3).
17. 29 C.F.R. 1908. 3(c).
18. Letter to Donald Whorton, M. D., Health Research Group, Washington, D. C., from George Guenther, Assistant Secretary of Labor for Occupational Safety and Health (October 3, 1971), p. 5.
19. P.L. 91–596, Section 24(a)(2).

Chapter 9

1. Remarks of President Nixon (December 29, 1970), reported in *Safety Standards*, U. S. Department of Labor, Vol. 20, No. 2, p. 4.

2. *Congressional Record* (November 17, 1970), S18340.
3. *Congressional Record* (March 30, 1971), E2485.
4. P.L. 91–596, Section 7(a)(1).
5. Reprinted in 1970 Senate Hearings, part 2, p. 1770.
6. Hearings on H.R. 10061, Departments of Labor and HEW appropriations for 1972, before the Senate Committee on Appropriations, 92nd Congress, 1st Session, part 3 (1971), p. 1496; Hearings on H.R. 10061, Departments of Labor and HEW appropriations for 1972, before a subcommittee of the House Committee on Appropriations, 92nd Congress, 1st Session, part 6 (1971), p. 447.
7. Hearings on Departments of Labor and HEW appropriations for 1973, before a subcommittee of the House Committee on Appropriations, 92nd Congress, 2nd Session (1972), p. 1182.
8. Interview with Sheldon Samuels, Industrial Union Department (IUD), AFL-CIO (November 3, 1971).
9. Letter to Mike Ryan, director for community and political action, Teamster Local 688, St. Louis, Missouri, from James Oser, P. E., assistant chief, Scientific Reference Service, National Institute for Occupational Safety and Health (NIOSH) (July 22, 1971).
10. Federal Bar Association, Bureau of National Affairs, Conference on OSHA, June 29, 1971.
11. Hearings on Departments of Labor and HEW Appropriations, *supra* note 6.
12. *Occupational Safety and Health Reporter*, publication of the Federal Bar Association, Bureau of National Affairs, No. 30 (November 25, 1971), p. 593; No. 29 (November 18, 1971), p. 586.
13. *Federal Register*, Vol. 36, No. 105 (May 29, 1971) pp. 10593–10595.
14. News Release, U. S. Department of Labor, Office of Information (April 28, 1971), p. 3.
15. *Congressional Record* (December 17, 1970), H11899.
16. *Federal Register*, Vol. 36, No. 157 (August 13, 1971), p. 15102.
17. *Federal Register*, Vol. 34, No. 96 (May 20, 1969), part II.
18. See 1968 House Hearings, pp. 736, 742–749.
19. Quoted in Franklin Wallick, "The Workplace Environment," in *Nixon and the Environment: The Politics of Devastation* (New York, 1972), p. 230.
20. Subcommittee on Intergovernmental Relations of the Committee on Government Operations, 92nd Congress, 1st Session, part 2 (1971), pp. 437–438.
21. P.L. 91–596, Section 8(g) (2).
22. 29 C.F.R. 1903.6.
23. Transcript of nomination testimony of George Guenther before the Senate Labor and Public Welfare Committee (April 2, 1971), p. 11 (copy on file with the committee).
24. Memorandum to all United Steelworkers of America district directors from I. W. Abel, president (October 27, 1971), p. 6.
25. Testimony of Jerome B. Gordon, 1968 Senate Hearings, p. 541.
26. Department of Labor, Press Release 72–381 (June 15, 1972).
27. "OSHA Non-Compliance Cited," *Journal of Commerce* (December 17, 1971), p. 2.
28. John Dizard, "OSHA Enforcement Pondered," *Journal of Commerce* (August 18, 1971), p. 2.

29. Mary Ellen Brennen, "The Hired Farmworker: The Consequences of His Powerlessness," paper submitted to Prof. Joseph A. Page (December, 1971).

30. Interview with George Guenther, Assistant Secretary for Occupational Safety and Health, in *Occupational Hazards* (October, 1971), p. 87.

31. Legislative History of the Occupational Safety and Health Act of 1970, prepared by the Subcommittee on Labor of the Senate Subcommittee on Labor and Public Welfare, June, 1971, p. 1067.

32. Daniel Berman, "Occupational Disease and Public Policy: The Case of Industrial Lead Poisoning in Missouri," paper prepared for the Department of Political Science, Washington University, St. Louis, Missouri (May 29, 1970), pp. 9, 13.

33. Missouri Interim Agreement under Section 18(h) of P.L. 91–596, on file at the U. S. Department of Labor, Office of State Plans, Washington, D. C.

34. Berman, *supra* note 32, p. 11.

35. Letter to Daniel Berman from Mr. Gerber, director, Environmental Health Section, Department of Public Health and Welfare, Missouri (April 2, 1971).

36. 1970 Senate Hearings, part 1, p. 609.

37. Interview with Richard Wilson, U. S. Department of Labor, Office of State Plans.

38. Notes taken by Study Group member at NACOSH meeting (September 24, 1971).

39. Memo to U. S. Chamber of Commerce Congressional Action Committee chairman and executives, from Hilton Davis, legislative action general manager (March 26, 1970), Fact Sheet No. 201a, p. 2.

40. Barry Brown, director, Michigan Department of Labor, "The State's Role under OSHA," speech presented at the National Safety Congress and Exposition, Chicago, Illinois, reprinted in *Congressional Record* (October 27, 1971), E213.

41. *Loc. cit.*

42. "OSHA: Labor's Strategists Apply Unrelenting Pressure," *Occupational Hazards* (October, 1971), p. 53.

43. Region 9 UAW meeting, New Jersey (November 23, 1971).

44. Press Release, UAW Public Relations Department (November 17, 1971), p. 2.

45. *Model Safety and Health Clauses*, UAW.

46. "Unions Spur OSHA Enforcement," *Journal of Commerce* (December 1, 1971), p. 2.

47. Letter to Mary-Win O'Brien from George W. Strugs, Jr., Social Security Department, UAW (January 4, 1972).

48. OCAW *Union News*, Vol. 27, No. 5 (July, 1971), p. 1.

49. Interview with Anthony Mazzocchi (November 8, 1971).

50. "On-the-Job Safety and Health Needs Get New-Breed Priority," UAW *Washington Report* (January 17, 1971), p. 2.

51. *Loc. cit.*

52. *Loc. cit.*

53. UE *Lansing Labor*, Vol. 27, No. 15 (November 19, 1971), p. 4.

54. Telephone interview with Larry Rubin (January 18, 1971).

55. Press Release, American Cancer Society News Service (November 3, 1971), p. 3.

56. *NLRB* v. *Knight Morley*, 251 F.2d 753 (6th Cir. 1958), and *NLRB* v. *Washington Aluminum*, 370 U. S. 9 (1962).
57. Charles Younglove, District 29, United Steelworkers of America, address before Second International Safety Conference, Chicago, Illinois (March 25–27, 1971), p. 35.
58. Ben Fischer, director, contract administration, United Steelworkers of America, address before Second International Safety Conference, Chicago, Illinois (March 25–27, 1971), p. 16.
59. Memorandum to all USWA district directors and staff representatives from I. W. Abel, president, on "Safety and Health Contract Language" (October 27, 1971), p. 23.
60. *Occupational Safety and Health Reporter*, Bureau of National Affairs, No. 27 (November 4, 1971), p. 517, citing address to the American Management Conference (October 27–29, 1971).
61. Telephone interview with Hal S. Breslow, labor relations officer, Continental Oil Co., Houston, Texas (November 23, 1971).
62. Steelworkers Memorandum, *supra* note 59, p. 13.
63. Ben Fischer, *supra* note 58.
64. Interview with Sheldon Samuels, Industrial Union Department, AFL-CIO (November 3, 1971).
65. Interview with George Taylor, AFL-CIO (November 8, 1971).
66. Sheldon Samuels, *supra* note 64.
67. "Nixon Uses AFL-CIO Convention to Air Political Appeal to Rank-and-File Workers," *Wall Street Journal* (November 22, 1971).
68. *Best's Safety Management Bulletin* announcing completed survey, cited in *Safety Management* (March, 1972).
69. "The Crushing Cost of Safety," *Dun's* (January, 1972).
70. *Loc. cit.*
71. *Accident Facts*, National Safety Council (1970 ed.), p. 25.
72. *Dun's, supra* note 69.
73. *Loc. cit.*
74. Elsie Carper, "Union Accuses Asbestos Plant of Selling Contaminated Bags," *Washington Post* (February 15, 1972), p. 3.
75. Interview with Anthony Mazzocchi (February 15, 1972).
76. Elsie Carper, *supra* note 74.
77. *New York Times* (June 4, 1972).
78. *Power Magazine*, June, 1971, p. 16.
79. See *Congressional Record* (March 8, 1972), S3627; (March 13, 1972), S3864; (March 20, 1972), H2264; (April 5, 1972), S5411.
80. See *ibid.* (May 30, 1972), E5867.
81. Reprinted, *ibid.* (July 17, 1972), E6801.
82. *Ibid.* (June 23, 1972), S10085.

Chapter 10

1. "Port Huron Death Toll May Exceed 22," *Washington Post* (December 13, 1971), p. 3; Jerry Flint, "22 Bodies Found in Blast in Tunnel," *New York Times* (December 13, 1971), p. 70.
2. "Lax Safety Precautions Reported at Tunnel Site," *Washington Post* (December 14, 1971), p. 2.
3. 1970 Senate Hearings, part 2, p. 1727.

4. Robert Sherrill, "How We Fail to Protect the Health of America's Workers," *Today's Health* (August, 1972), p. 65.
5. "The Manpower Shortage: It Will Get Worse Before It Gets Better," and "Youth: Not Turned On by a Career in Safety," *Occupational Hazards* (December, 1971), pp. 34, 36.
6. H.R. 14816, Section 6(a) (2), 1968 House Hearings, p. 3.

Index